Psoralen DNA Photobiology

Volume II

Editor

Francis P. Gasparro, Ph.D.
Research Scientist
Department of Dermatology
Yale University School of Medicine
New Haven, Connecticut

CRC Press
Taylor & Francis Group
Boca Raton London New York

CRC Press is an imprint of the
Taylor & Francis Group, an **informa** business

First published 1988 by CRC Press
Taylor & Francis Group
6000 Broken Sound Parkway NW, Suite 300
Boca Raton, FL 33487-2742

Reissued 2018 by CRC Press

Library of Congress Cataloging-in-Publication Data

Psoralen DNA photobiology/editor, Francis P. Gasparro.
      p.   cm.
   Bibliography: p.
   Includes index.
   ISBN 0-8493-4379-8. ISBN 0-8493-4380-1
   1. Psoralens--Physiological effect. 2. DNA. 3. Photobiology.
4. DNA probes. I. Gasparro, Francis P.
QP801.P686P76 1988
612'.01444--dc19   88-10437

A Library of Congress record exists under LC control number: 88010437

ISBN 13: 978-1-315-89702-8 (hbk)
ISBN 13: 978-1-351-07612-8 (ebk)

Visit the Taylor & Francis Web site at http://www.taylorandfrancis.com and the
CRC Press Web site at http://www.crcpress.com

PREFACE

In one form or another psoralens have been in use dating back to biblical times for the treatment of depigmented patches of skin. However, it has only been in the past 40 years that the structure and function of psoralens have been elucidated. Although several volumes have been published on photobiology and photomedicine, no one volume has ever been devoted to the psoralen photobiology. In this book we focus on the properties and uses of photoactivated psoralens.

In these volumes the various aspects of psoralens are presented in a review of the field as it stands in mid 1986. In retrospect, we may find that this particular time was crucial in the development of new therapeutic modalities as many of the applications of modern molecular biology are beginning to impact on the practice of medicine. This book was written with two purposes in mind. First, to serve as an update (the last collective review of the field was in 1982). Second, it is hoped that newcomers to the fields of photobiology and photomedicine — both scientists and clinicians — would find it a useful introduction.

The topics in this book range from basic science to medical applications. Also, a brief summary of the development of the use of psoralens in the U.S. has been authored by A. Lerner (Yale University) one of the pioneers in the field of psoralen photobiology. In Chapter 1 elementary aspects of psoralen-DNA interaction are presented. A newcomer to psoralen photobiology should be able to read this chapter and then comprehend current papers in the research literature. For the photobiologist, this chapter collects diverse thermodynamic and photochemical data. G. Rodighiero and F. Dall'Acqua, current members of the Italian group at Padua who have contributed so greatly to the initial development of psoralen photochemistry, have authored a chapter on angelicins (in collaboration with D. Averbeck of the Curie Institute in Paris). Data in this chapter describe the effects of various chemical substituents on angelicin photochemical properties and show that these compounds, although unable to form crosslinks, possess potent therapeutic properties.

In Chapter 3, R. W. Midden describes the effects of photoactivated psoralens on cellular components other than nucleic acids. Certainly, DNA based pairs provide an exquisitely receptive site for psoralen intercalation; however, other biomolecules (proteins and membrane components) have been shown to undergo photoreactions with psoralens. Whether these non-nucleic acid reactions have photobiological effects remains to be determined.

Methods for the detection of psoralen photoadducts provide the opportunity to determine whether a correlation exists between the number of adducts formed and therapeutic efficacy. In Chapter 4, R. Santella (Columbia University) describes the use of highly specific monoclonal antibodies that we developed specifically for that purpose. In addition the availability of the antibodies will eventually permit the determination of the in vivo pattern of photoadduct formation in human DNA.

Psoralens have also proved useful as probes of nucleic acid structure and function. In Chapter 5, P. W. Wollenzein (St. Louis University) describes experiments that give specific information about the spatial arrangement of nucleic acids and their interaction with nucleosomal proteins.

As with many other chemotherapy agents directed against DNA, psoralens plus UVA have been shown to induce genotoxic effects. In Chapter 6, W. A. Saffran reviews the mutagenic and recombinogenic effects of psoralen plus UVA including a dissection of the effects of monoadducts and cross-links.

Psoralen phototherapies are thought to be effective because the DNA damage impedes proliferating cells. However, normal cells are also affected. Persisting unrepaired damage in normal cells or misrepaired psoralen photoadducts leads to mutagenic activity. In Chapter 7, P. D. Hanawalt and C. A. Smith (Stanford University) review the repair of psoralen photoadducts in model systems.

In Chapter 8, R. Knobler and H. Honigsmann (University of Vienna) and R. Edelson (Yale University) describe the use of psoralens for the treatment of a variety of dermatological conditions. In addition, a new photochemotherapy for cutaneous T cell lymphoma is described in which a patients blood is leukapheresed and then irradiated extracorporeally prior to reinfusion.

Although these volumes review the field of psoralen-DNA photobiology, they also highlight challenges that remain to be overcome. First, the complete pattern of in vivo photoadducts formation and the respective role(s) of monoadducts and cross-links remain to be determined. Second, the in vivo effects of non-nucleic acid reactions and their role may explain the effectiveness of psoralens plus UVA in the treatment of vitiligo. Third, the availability of specific monoclonal antibodies will permit the characterization of adduct repair and persistence in human cells. Fourth, the conjugation of psoralens to receptor-seeking biomolecules (monoclonal antibodies, transferrin, insulin, etc.) may turn a cellularly nonspecific drug into an agent which could be delivered to a selected subpopulation of cells. Finally, the question remains of how to evaluate new psoralens. Traditionally psoralens have been tested for potential clinical efficacy on the basis of their ability to photosensitize skin. Ample data now illustrate that this method may bypass some potentially valuable derivatives (see Chapter 2). Furthermore, a standard set of assay conditions should be followed. Too often, different investigators have used vastly differ concentrations of the drug (up to micrograms per milliliter) or doses of UVA light. At this point, it is clear that a typical psoralen blood level falls within the range of 50 to 150 ng/mℓ. Perhaps 100 ng/mℓ should be a standard test level for in vitro experiments. With regard to light doses, it is important to realize that although the determination of impinging doses are easily made, the amount of light actually reaching target cells (skin in PUVA and lymphocytes in photopheresis) is vastly attenuated by passage through upper levels of the skin or red blood cells, respectively. Thus, for in vitro studies, clinically relevant doses are probably in the range of 0.5 to 2 J/cm^2.

The editor gratefully acknowledges the invaluable assistance of many colleagues at Yale University, especially Drs. Roger Kahn, Peter W. Heald, and Paula M. Bevilacqua. John Battista and Lawrence Weisman, members of his laboratory, also assisted in the preparation of the manuscript for Chapter 1. The editor also thanks Dr. W. R. Midden who commented on early drafts of Chapter 1. The rationale for much of my research on 8-MOP-DNA interactions has evolved while developing a new photochemotherapy for the treatment of cutaneous T cell lymphoma (photophoresis). Much of this work would not ahve been possible without the generous support of Johnson and Johnson, both as a corporate institution and as a collection of hard-working, dedicated individuals. Kyu Ho Lee, John MacLean, George Sillup, Joseph Garro, William Lindemann, and John Taylor are Therakos scientists, engineers, and managers each of whom made specific and concrete contributions. In additon, the dedication of the two photophoresis nurses, Glynis McKiernan and Inger Christensen who often drew blood samples for 8-MOP levels and photoadduct determination is gratefully acknowledged. Additional support during the past year from Therakos, Inc. (a Johnson and Johnson subsidiary) greatly fascilitated the preparation of these volumes.

The research summarized or referred to in Chapter 1 has been supported by a New Investigator Award from the NIH (AM-37629) and a grant from the Matheson Foundation. Support and encouragement during the preparation of these volumes from Dr. Richard L. Edelson, who first introduced the editor to the field of psoralen photobiology, is gratefully acknowledged.

Francis P. Gasparro, Ph.D.
Yale Universlity
April 13, 1987

In my mother's memory who
passed away while these volumes were in preparation.

THE EDITOR

Francis P. Gasparro, Ph.D., is currently a research scientist with the Department of Dermatology at the Yale Unversity School of Medicine in New Haven, Connecticut. Dr. Gasparro received his Bachelor's degree in Chemistry from Villanova University in 1966 and his Ph.D. from Princeton University in 1971. Before reaching his current position, he was a research associate with the Department of Biochemical Sciences at Princeton following graduation and then moved to Columbia as a staff associate and later as associate research scientist in the Department of Dermatology.

Dr. Gasparro has taught and lectured at many colleges and universities and has been the recipient of numerous research grants. He has written many journal articles, abstracts, and papers and is a member of the American Society of Photobiology. Dr. Gasparro's research interests include psoralen DNA photochemistry, immunoassay for psoralen DNA photoadducts, and photochemistry of other molecules of biological and medical interest.

CONTRIBUTORS

Dietrich Averbeck
Department of Radiation Biology
Curie Institute
Paris, France

Francesco Dall'Acqua
Department of Pharmaceutical Sciences
University of Padova
Padova, Italy

Richard L. Edelson, M.D.
Professor and Chairman
Department of Dermatology
Yale University School of Medicine
New Haven, Connecticut

Francis P. Gasparro, Ph.D.
Research Scientist
Department of Dermatology
Yale University School of Medicine
New Haven, Connecticut

Herbert Honigsman, M.D.
Professor
Department of Dermatology
University of Vienna
Vienna, Austria

Robert M. Knobler, M.D.
Research Scientist
Department of Dermatology
University of Vienna
Vienna, Austria

Debra L. Laskin, Ph.D.
Assistant Professor
Department of Pharmacology and
 Toxicology
Rutgers University
Piscataway, New Jersey

Jeffrey D. Laskin, Ph.D.
Associate Professor
Department of Environmental and
 Community Medicine
Robert Wood Johnson Medical School
Piscataway, New Jersey

Aaron B. Lerner, M.D., Ph.D.
Professor
Department of Dermatology
Yale University School of Medicine
New Haven, Connecticut

W. Robert Midden, Ph.D.
Assistant Professor
Department of Chemistry
Bowling Green State University
Bowling Green, Ohio

Giovanni Rodighiero
Department of Pharmaceutical Sciences
University of Padova
Padova, Italy

Wilma Saffran, Ph.D.
Associate Research Scientist
Department of Human Genetics
Columbia University
New York, New York

Regina Santella, Ph.D.
Assistant Professor
Institute of Cancer Research
Columbia University
New York, New York

Charles A. Smith, Ph.D.
Department of Biological Sciences
Stanford University
Stanford, California

Paul L. Wollenzien, Ph.D.
E. A. Doisy Department
St. Louis University Medical School
St. Louis, Missouri

TABLE OF CONTENTS
Volume I

Volume II

Chapter 4

CHEMICAL MECHANISMS OF THE BIOEFFECTS OF FUROCOUMARINS:
THE ROLE OF REACTIONS WITH PROTEINS, LIPIDS, AND OTHER
CELLULAR CONSTITUENTS

W. Robert Midden

TABLE OF CONTENTS

I. INTRODUCTION

This chapter will examine the chemical reactions of furocoumarins that could be responsible for their biological effects. Special attention will be given to those reactions that do not involve DNA, since the role of DNA reactions in furocoumarin chemistry and biology has been covered many times in the past[1-22] and in other chapters of this monograph (Chapters 1, 2, and 6). In particular, a novel mechanism for the antiproliferative effects of psoralens which are important in the therapy of psoriasis will be proposed in this chapter. This mechanism involves alteration of the signaling pathways of eukaryotic cells by formation of covalent adducts between furocoumarins and unsaturated fatty acids. Even though non-DNA reactions will be emphasized, this review will nevertheless draw on many examples of furocoumarin-DNA chemistry because of the important role these concepts have played in the development of our understanding of mechanisms of furocoumarin actions.

A. History

Furocoumarins have been used for centuries in the form of crude and purified plant extracts to treat vitiligo (leukoderma).[23-26] Ingestion or contact with the plants that are used for this therapy or with a number of other plants including parsnip, parsley, and celery can cause severe photosensitization termed phytodermatitis.[27-29] Progress in understanding the chemical reactions responsible for the photosensitization caused by these plants came slowly. In 1834 Kalbruner reported the first isolation of the photosensitizer 5-methoxypsoralen (5-MOP) from bergamot oil,[30] but it was not until 1931 that Phyladelphy demonstrated that sunlight was a necessary component in the action of these compounds.[31]

B. Psoriasis Therapy

In the early 1970s a new therapy for the treatment of psoriasis was proposed based on the use of these plant derivatives.[32-33] The treatment involved oral ingestion or topical application of 8-methoxypsoralen (8-MOP) followed by illumination of the skin with near UV light (UVA). Since that time the effectiveness of these drugs for psoriasis has been confirmed,[34-36] and their efficacy has been demonstrated for the treatment of a variety of other skin diseases[47] including mycosis fungoides,[48-51] cutaneous lymphomas,[52] pustular psoriasis,[53] pustular palmaris and plantaris,[54] and actinic prurigo.[55]

C. Bioeffects

Besides the therapeutic benefits, the effects of furocoumarin plus UVA treatment include phototoxicity exhibited as skin erythema, edema, and vesiculation;[56-58] hyperpigmentation;[59-74] ocular cataracts[65-70] (although the risk of cataract formation has been questioned[71-74] and should be easily prevented with adequate eye protection); and carcinogenesis.[75-84] Consistent with their carcinogenicity, furocoumarins plus UVA can also cause mutations in prokaryotic and eukaryotic cells and viruses.[85-89] While the latest epidemiological results tend to suggest that the carcinogenicity of PUVA in humans is weak, it is worth noting that ionizing radiation was considered safe as long as 15 years after its first use to treat psoriasis.[83,90] It is important to consider the long latent period for tumors in humans when evaluating the carcinogenicity of psoralen plus UV light (PUVA) therapy.[91] The combination of these harmful and beneficial effects has stimulated interest in the chemical mechanisms of the actions of furocoumarins in the hope that a means would be discovered for designing a drug with no harmful side effects, especially one that would lack carcinogenicity.

D. General Chemistry of Furocoumarins

Furocoumarins are compounds which contain a furan ring fused to a coumarin molecule. Furocoumarins which have the fusion at the 2,3 bond of the furan and the 6,7 bond of the coumarin are considered linear furocoumarins and are commonly called psoralens. Fusion at the 2,3 bond of the furan and the 7,8 bond of the coumarin are considered angular furocoumarins and are commonly called angelicins. The fundamental photochemistry of furocoumarins, including the primary photoprocesses such as formation of the singlet and triplet states, fluorescence, and phosphorescence, is covered in Chapter 1 and therefore will not be discussed in detail here. The basic tenets of photochemistry of organic molecules, in general, is available in many photochemistry texts[92-93] and in the excellent summary by Lamola that emphasizes those principles that are especially relevant to the photochemistry of agents that damage biomolecules.[94] Various substituents such as methyl, ethyl, methoxy, ethoxy, carbethoxy, hydroxyl, and amino can be attached to the furocoumarin ring to alter its photochemistry and biological effects, and these substituents play an important role in determining the chemical and biological characteristics of these drugs.

The effects of psoralens certainly depend on factors at many different levels: rates of absorption into the organism, rates and paths of transport throughout the organism, rates of chemical transformation, rates of penetration into different cell types, and rates of excretion or elimination as well as the effective light dose at the various cellular targets. This chapter will consider primarily the identity of the molecular targets within cells and the chemical reactions by which these targets are modified rather than the gross biological or physiological factors which undoubtedly also play an important role in determining the net effects.

II. PHOTOADDITION TO NUCLEIC ACIDS

One of the first biochemical reactions of furocoumarins that was identified in the search for the molecular target within cells was the light-induced covalent binding of psoralen to nucleic acids.[95-99] A large number of papers confirming this observation and examining many details of this reaction were subsequently published. Besides cycloaddition to pyrimidines, recent reports indicate that furocoumarins may also bind to purines.[100-102]

A. Viral Inactivation

While it is easy to demonstrate that furocoumarins efficiently bind to DNA, it is still necessary to show that this reaction is actually responsible for biological effects. Some biological activities of furocoumarins, such as the inactivation[95,103-106] and the induction of viruses,[107] have been relatively easy to conclusively attribute to modification of nucleic acids. Evidence that DNA reactions are capable of causing inactivation was also obtained by Colombo with sea urchin sperm.[108] In fact, the ability of psoralens to react with DNA is considered so characteristic that the inability of psoralens to cause inactivation has been used as evidence of the lack of nucleic acid in the scrapie agent.[109]

B. Mutagenicity

The mutagenicity[85-89,110] of furocoumarins in bacteria and eukaryotic cells is also easily attributed to modification of DNA, since nearly all mutagens have been found to modify DNA in some way.[111-112]

C. DNA Repair Deficiency

A number of observations strongly support the hypothesis that modification of DNA is responsible for the lethal effect of furocoumarins in microorganisms as well as virus and sperm. For instance, organisms that are deficient in DNA repair are often more easily killed by the photochemical action of psoralens. In 1975 Averbeck and Moustacchi found that Saccharomyces deficient in either excision or recombinational DNA repair are more sensitive than wild type cells to PUVA (8-MOP).[113] The double mutant was more sensitive than the single mutants. The role of DNA damage in furocoumarin cytotoxicity is also supported by the observation that furocoumarins have been found to inhibit DNA synthesis more than RNA or protein synthesis in a variety of cells.[16,103,114-121] The importance of DNA repair for reducing the cytotoxicity of psoralens has also been demonstrated in human cells. Cells from patients with xeroderma pigmentosum, congenitally defective in DNA excision repair, were found to be more sensitive to killing by 8-MOP than normal cells even though DNA synthesis was inhibited to the same extent in both types of cells.[114] Repair of 8-MOP UV-damaged DNA could be demonstrated in normal cells but not in those from xeroderma pigmentosum. It is interesting to note, however, that treatment with 8-MOP plus UVA light resulted in less repair synthesis of DNA relative to the inhibition of replication synthesis of DNA than treatment with UVC light alone. This may be due to greater inhibition of DNA repair enzymes by 8-MOP than by UVC.

D. Is DNA the Only Target?

It is easy to see why DNA binding most often has been cited as the critical reaction responsible for the bioeffects of furocoumarins.[6,7] However, it is important to consider the variety of biological effects which these drugs can exhibit and to consider the possibility that some effects may arise from other types of molecular lesions. It is possible that a given type of furocoumarin biological activity could arise from a combination of chemical events and that each type of chemical event could be a necessary or sufficient condition for expression of the effect. Besides cytotoxicity and mutagenicity in microorganisms, the biological effects of furocoumarins in humans include inhibition of hyperplasia (therapy of psoriasis), stimulation of pigmentation (enhanced tanning and therapy of vitiligo), erythema, inflammation and edema (symptoms characteristic of a wide variety of toxic agents), and cataractogenesis. While it is likely that furocoumarin photomutagenesis is due to direct modification of DNA, it is important to remember that the other biological effects have not been *proven* to result from the modification to DNA.

A number of observations are consistent with the possibility that reactions with targets other than DNA may account for some furocoumarin bioeffects. Most studies of the intracellular localization of psoralens have found relatively little psoralen attached to DNA after UVA illumination. Pathak and Kramer found only 0.18% of topically applied psoralen bound to the nuclear fraction of guinea pig skin even under conditions which strongly favored binding of psoralen to DNA (topical application of a relatively large amount of 8-MOP to skin that had been stripped of the stratum corneum and a relatively large light dose).[127] This represents only one in every 70,000 bases modified with a dose of 8-MOP and UVA that caused a considerable phototoxic response. Most (99%) of the radioactivity did not penetrate the surface of the skin. Of the portion that penetrated to the epidermis, 82% was found in the nonsedimenting (soluble) fraction representing the cytoplasm. Of the radioactivity in this material, 99.5% was recovered in a low-molecular-weight fraction isolated by gel chromatography and therefore could correspond to unreacted psoralen or psoralen bound to low-molecular-weight molecules such as lipids or amino acids. Only 0.5% of the cytoplasmic radioactivity was found bound to soluble proteins, and this binding could be accounted for by association of psoralen photolysis products with the proteins — a process which does not require light and may involve noncovalent binding.[99] Another study of the labeling of various cellular fractions by tritiated 8-MOP led some investigators to conclude that administration of therapeutic levels of 8-MOP to rats leads to binding to proteins at much higher levels than nucleic acids.[128]

Significant binding of furocoumarins to targets other than DNA could explain the apparent lysosome lysis observed by Fredericksen and Hearst in HeLa and *Drosophila* cells. Their treatment of cells with psoralens and UVA caused an increased rate of degradation of RNA which was attributed to release of lysosomal enzymes.[125] Lysosome lysis implies that damage to proteins or lipids in the lysosome membrane occurred. Furocoumarins were observed to bind to proteins in this study, but it was difficult to determine whether binding to proteins required light or involved covalent bonds because of the high affinity of the dark binding. Lysosome lysis was also suggested by increased proteolytic activity observed by Meffert and colleagues.[126] Whether reactions with proteins, lipids, or some other molecular target are responsible for this lysis is unknown, but it is reasonably certain that it was not due to reactions with nucleic acids.

Measurements of the levels of formation of psoralen-DNA adducts using fluorescence-labeled specific immune serum shows that psoralen-DNA adducts are produced at lower levels in living organisms than in cell culture or isolated DNA. The levels of psoralen-DNA adducts were undetectable until doses of the psoralen and UVA were much higher than that required to cause a severe cytotoxic response.[129] These results led the authors to point out that "the question arises whether DNA-psoralen-photoadducts which are formed at so low a number in epidermal nuclei under PUVA conditions can be considered as the only or principal cause of inhibition of epidermal cell proliferation".

Others have also found that binding to other cellular components was as extensive if not greater than binding to nucleic acids. Recent evidence obtained by Laskin et al. has demonstrated specific, saturable, high-affinity binding sites for psoralens on the membranes of cells.[130] This group observed this binding in HeLa cells as well as four other human cell lines and five mouse cell lines, suggesting that this is a general characteristic. Bredberg et al.[131] and Bertaux et al.[132] found roughly equal localization of ^{3}H-labeled 8-MOP in the cytoplasm and the nucleus of cells by autoradiography. The latter group found that keratin, collagen, and lipoproteins were labeled and concluded that proteins as well as DNA could be targets of furocoumarins. Fluorescence microscopy by Meffert et al. shows binding in the stratum corneum, in cellular membranes, and in the cytoplasm of the cells of the malphghian layer.[133] Electron microscopy of guinea pig skin also provided evidence suggesting that photobinding to other targets besides DNA could be occurring.[134] The rapid dose-dependent reduction in the rate of oxygen consumption in rat brain and liver homogenates upon treatment with 8-MOP and light, observed by Ali and Agarwala, is most easily attributed to protein damage rather than nucleic acids, since the reductions occurred before DNA damage would have had time to be expressed.[135] The fluorescence increase in hair and epidermis observed as a result of PUVA treatment also suggests that psoralen binding to proteins can occur.[136] Protein damage may be especially critical in the formation of ocular cataracts,[65,67,122,123] since the turnover rate of ocular proteins tends to be much slower than the protein turnover rates for other tissues.[66,68,70,124]

This chapter will examine the photochemistry of furocoumarins with cellular constituents other than DNA to evaluate the role these other reactions might play in furocoumarin biological activity. However, since the photocycloaddition to DNA has been most thoroughly studied, we will begin with a brief review of the general characteristics of this reaction to identify any features that might be relevant to reactions with other cellular components. Details of this chemistry are available in Chapter 1.

E. General Chemistry of the Cycloaddition Reaction

The furocoumarin-induced photochemical modification of DNA by furocoumarins involves the formation of a furocoumarin-DNA adduct. This reaction is a 2 + 2 cycloaddition forming a cyclobutane ring joining the furocoumarin and the 5,6 double bond of the pyrimidine. For psoralen and many of its derivatives, reaction can occur at either the 3,4 bond of the pyrone ring or the 4′,5′ bond of the furan ring or both. Furocoumarins, such as psoralen, that are capable of reacting at both the 3,4 and the 4′,5′ bonds are termed bifunctional. In double-stranded DNA, interstrand cross-links can be formed with bifunctional psoralens, and this ability to form DNA cross-links has been exploited for investigating the structure of nucleic acids.[3,5] Angular furocoumarins, such as the angelicins, have not been observed to cross-link DNA, apparently due to the unfavorable geometry. Psoralens, such as 3-carbethoxypsoralen (3-CPs), that have the reaction blocked at one end of the molecule also form only monoadducts with DNA.

1. Stereochemistry of Cycloaddition

The addition of many psoralens to native double-stranded DNA is a relatively efficient photochemical reaction with maximum quantum yields as high as 0.08.[5] Furocoumarins also add to free pyrimidines in aqueous solution, but the reaction is much slower; in addition, the product distribution is different with free pyrimidines than with double-stranded DNA. Molecular orbital calculations indicate that the furocoumarin 3,4 bond should be the most reactive position in psoralen, since this is where the greatest electron density is located[137-139] and this is the bond that is most reactive toward free pyrimidines in solution. That this reaction occurs via the furocoumarin triplet state is supported by the inhibition by oxygen and paramagnetic ions.[140] However, in double-stranded DNA, the 4′,5′ adduct forms most rapidly.[141-143] This difference in stereochemistry has been attributed to the geometry of the furocoumarin with respect to the pyrimidine when it is intercalated in DNA, but it may also be due to the fact that when furocoumarins are intercalated in DNA, reaction via the singlet state can occur due to the close association of the pyrimidine and the furocoumarin. The singlet state is predicted to react preferentially at the 4′,5′ bond.[137] Indeed, when psoralen or 4′-aminomethyl-4,5′,8-trimethylpsoralen (AMT) are bound to DNA, the triplet state cannot be detected,[144] most likely due to depletion of the singlet as the result of physical or chemical quenching by DNA.

III. REACTIONS WITH LIPIDS AND MEMBRANES

A. Photoaddition to Unsaturated Fatty Acids

However, pyrimidines are not the only compounds which undergo photocycloaddition with furocoumarins. As stated in a text on organic photochemistry, ''The formation of dimers on irradiation of compounds that contain olefinic bonds is one of the oldest photochemical reactions known, and a vast number of examples have been reported.''[145] Besides cycloaddition to pyrimidines, furocoumarins also undergo photocycloaddition with themselves to form dimers and to simple olefins to form cyclobutyl adducts.[146] Consistent with the ability to form adducts with olefins, the efficient formation of covalent adducts between unsaturated fatty acids and psoralens has been observed.[147-149] When simple unsaturated fatty acids such

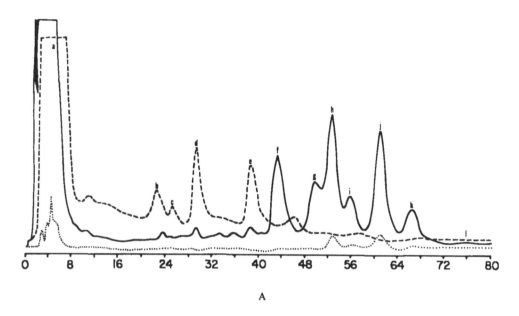

A

FIGURE 1. Oleic acid methyl ester (OAME) 10^{-3} M or stearic acid (SA) 10^{-3} M and 2×10^{-4} M tmPso in methanol-water 2:1 (v/v) were illuminated for 2 hr at a distance of 2 cm from four General Electric F 1578 BLB fluorescent bulbs with an intensity at the sample of 6×10^{-9} Einsteins/cm^2/sec (20 J/m^2/sec) and maximum emission at 360 nm. The products were analyzed without derivatization by reverse-phase HPLC using a (25 $\times$ 0.46-cm) Dupont Zorbax ODS column and eluting with acetonitriler 89:11 (v/v) at 1.0 mℓ/min in a Kratos FS950 fluorimeter with a 254-nm interference excitation filter, and a 320-nm cutoff emission filter was used to monitor the relative fluorescence of the eluant. Absorbance at 250 nm and 215 nm was monitored using a Varian UV-50 detector. The chromatograms are plotted as relative absorbance or fluorescence against retention volume (milliliters). (A) OAME-tmPso illuminated 2 hr (---, relative fluorescence emission at wavelengths greater than 320 nm, excitation at 254 nm; ———, absorbance at 215 nm; ···, absorbance at 250 nm). (B) Controls: illuminated SA-tmPso (---, relative fluorescence emission at wavelengths greater than 320 nm, excitation at 254 nm; ———, absorbance at 215 nm) and OAME-tmPso mixed for 2 hr in the dark (-···-, absorbance at 215 nm; -·-·-, absorbance at 254 nm; ···, relative fluorescence emission at wavelengths greater than 320 nm, excitation at 254 nm). (Peak identities: a, tmPso; f, OAME; b, c, d, e, g, h, i, j, k, and l, photoproducts.)

as oleic or linoleic acid are mixed with the common furocoumarins, 8-MOP, or 4,5′,8-trimethylpsoralen (TMP) and irradiated with standard fluorescent black lights (maximum emission at 360 nm, 20 J/m^2/sec), products can be detected within 15 min. The products exhibit chromatographic behavior that is much different from that of the typical products of fatty acid oxidation in that they have longer retention times on a reverse-phase HPLC column than does the parent fatty acid. This suggests that these compounds are more hydrophobic than the parent fatty acids, whereas nearly all fatty acid oxidation products are less hydrophobic due to the addition of oxygen-containing functional groups.

Analysis of a mixture of TMP and oleic acid methyl ester (OAME) by high pressure liquid chromatography (HPLC) after irradiation for 2 hr indicated that at least eight products were formed, as shown in Figure 1. Similar products were formed when mixtures of TMP or 8-MOP and other unsaturated fatty acids, such as linoleic, linolenic, and arachidonic acid, were irradiated. None of these products was observed in unirradiated mixtures of TMP and OAME or in irradiated mixtures of TMP or 8-MOP and saturated fatty acids such as stearic acid.

Isolation of the products formed from ³H-labeled TMP and ¹⁴C-labeled oleic acid and analysis by dual label scintillation counting shows that the products of this reaction contain both isotope labels in stoichiometric ratios. Fast atom bombardment mass spectrometry (Figure 2) has demonstrated that the molecular weight of the adduct is the same as the sum

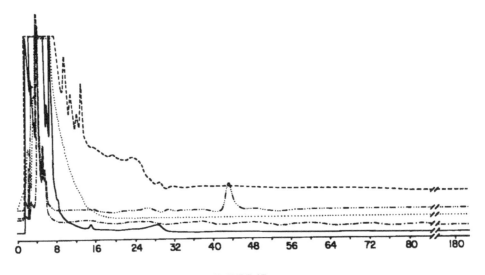

FIGURE 1B.

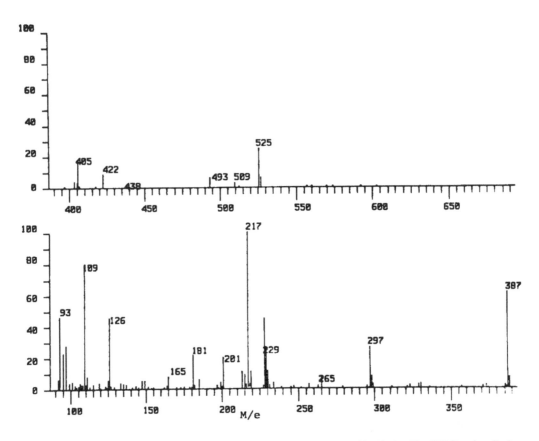

FIGURE 2. Fast atom bombardment mass spectrum for the OAME-tmPso adduct isolated by HPLC as described
in the legend of Figure 2, except using a 10-mm × 50-cm column of 7-μm C-18 Nucleosil derivatized silica and
eluting at 5 mℓ/min. The spectrum was determined in chloroform-thioglycerol matrix at the Middle Atlantic Mass
Spectrometer Laboratory. (Peak assignments: 525, MH^+; 493, MH^+-CH_3OH; 297, OAME·H^+-CH_3OH; 229,
tmPso·H^+.)

of the molecular weights of OAME and TMP. In water-methanol mixtures, the reaction between TMP and unsaturated fatty acids does not appear to involve significant oxidation of the fatty acids, in contrast to reactions sensitized by common photodynamic dyes such as rose bengal or methylene blue. A competitive kinetics technique[150,151] has shown that the disappearance of the fatty acid and the formation of the furocoumarin-fatty acid adduct does not involve singlet oxygen as an intermediate even though some singlet oxygen is formed. This is probably due to the relatively slow rate of reaction of unsaturated fatty acids with singlet oxygen and the relatively rapid rate of cycloaddition. Other substrates that do not undergo cycloaddition to psoralens but are rapidly oxidized by singlet oxygen, such as histidine, undergo nearly exclusive oxidation via singlet oxygen in the TMP- or 8-MOP-sensitized reaction. Which type of reaction occurs, cycloaddition or oxidation, depends on the relative rates of these two reactions for each substrate under the specific reaction conditions.

In fact, psoralen-fatty acid adducts are formed even in solutions that have been carefully purged of oxygen with nitrogen or argon from which traces of oxygen had been carefully removed.[152] Even in the presence of oxygen, no formation of fatty acid peroxides was detected by the starch-iodide test. It is notable that some of the bioeffects of psoralens are also found to be oxygen independent.[16,59,153-155]

Four diastereomeric products could be formed by 2 + 2 cycloaddition of the fatty acid to the 3,4- and four more diastereomers by addition to the 4',5'- bond of the furocoumarin. These products correspond to addition on one side or the other of the furocoumarin with the carboxyl group of the fatty acid either to the right or the left as illustrated in Figure 3 for the four 3',4' adducts. Based on the characteristics of other cycloaddition products of furocoumarins, products of addition at the 3,4 bond are typically nonfluorescent, while products of addition at the 4',5' bond are typically fluorescent. Four of the products formed in the reaction of TMP with OAME appear to be fluorescent, while four others appear to lack fluroescence on the basis of measurements made with a fluorescence emission detector of the effluent from the HPLC column (Figure 1). The nonfluorescent products may correspond to adducts of the fatty acid with the 3,4 bond of TMP, while the fluorescent products may correspond to addition at the 4',5' bond. Additional products that are apparent in the HPLC analysis may correspond to addition products in which the pyrone ring has undergone opening by hydrolysis. Such products have been previously observed with furocoumarin pyrimidine adducts.[156,157] Based on absorbance at 215 nm, the yield of the nonfluorescent compounds appears to be higher than the yield of the fluorescent products. Subsequent studies have shown that under most conditions, no fluorescent adducts are formed in detectable quantities. The nature of the fluorescent products previously observed has not been established. This would be consistent with the more rapid formation of 3,4 adducts in cycloadditions of psoralens in solution and the prediction of this behavior on the basis of molecular orbital calculations.[137,138] The more rapid formation of 4',5' adducts in double-stranded DNA may be due to geometrical factors as the result of intercalation of the psoralen into DNA that favor formation of the 4',5' adduct, or they may be due to reaction via the excited singlet psoralen made possible by the close association of the two reactants. While the singlet state may react most rapidly at the 4',5' position of the furocoumarin, the triplet state is predicted to react most rapidly at the 3,4 position. Reactions of unassociated molecules in solution may be more likely via the triplet state. The singlet state may decay too rapidly for efficient trapping by molecules freely diffusing in solution. However, the short-lived singlet state may be efficiently trapped when it is formed by excitation of furocoumarins intercalated in close association with pyrimidines in double-stranded DNA.

The formation of adducts between TMP and OAME proceeds with a quantum yield in methanol-water or ethanol solutions as high as 0.02, comparable to the quantum yield of the photoreaction of furocoumarins with DNA, suggesting that binding to fatty acids in cells may occur at least as fast as binding to DNA. Which reaction is favored, addition to nucleic

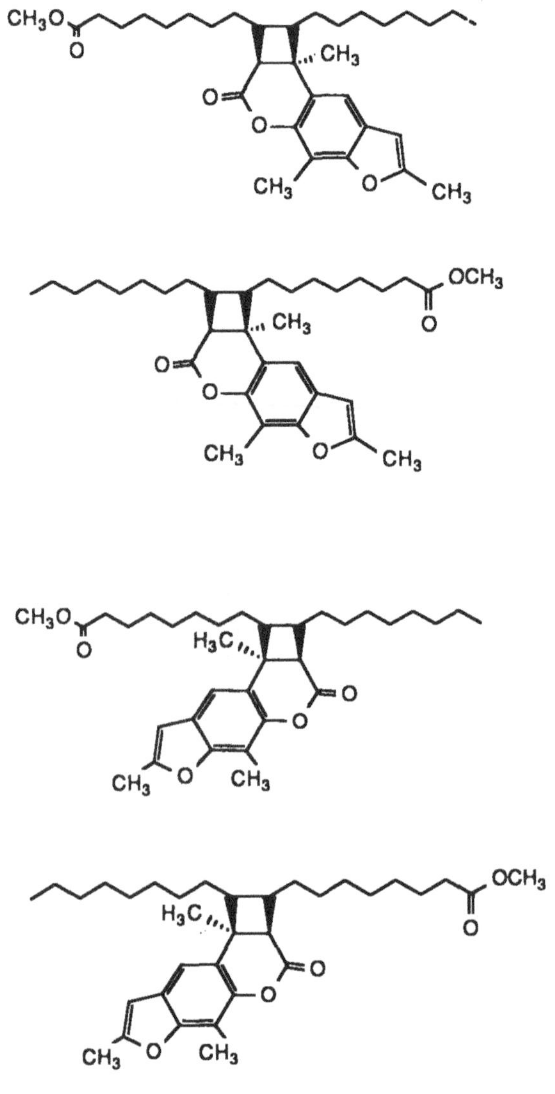

FIGURE 3.   Possible chemical structures of the four diastereomers which
may form by a 2 + 2 cycloaddition of the 3,4 bond of tmPso to OAME.

acids or to lipids, may depend on the extent of noncovalent association of the furocoumarins
with the lipid membranes and nucleic acids. More hydrophobic furocoumarins, such as TMP,
may be expected to associate more readily with the relatively hydrophobic environment of
lipid membranes, while planar, positively charged furocoumarins, such as AMT, might be
expected to bind strongly to DNA. The specific, high-affinity binding sites observed by
Laskin et al. in several lines of animal cells appear to be covalently modified upon irradiation
with UVA light.[130] Further investigation is needed to determine whether these binding sites
involve lipids or proteins.

B. Alteration of Signaling Pathways
At first glance it might not be evident how covalent attachment of furocoumarins to what

serve primarily as structural molecules in a cell could profoundly alter the rates of cell proliferation and metabolism and thus account for the observed bioeffects of furocoumarins. However, recent research has demonstrated a very important role for some lipids in mediating hormonal signals and other processes crucial to cell regulatory activities. Among the signaling processes that are known to be coupled to activation of plasma membrane receptors are the stimulation and inhibition of adenylate cyclase, activation of protein kinase activities, and the opening of ion channels which lead to a change in membrane polarization or to a change in the intracellular concentration of ions such as Ca^{2+}.[158] These processes play a crucial role in determining the rates of cell division and the regulation of cell metabolism. Phospholipase C-catalyzed hydrolysis of phosphatidylinositol-4,5-bisphosphate, a quantitatively minor membrane phospholid, yields *sn*-1,2-diacylglycerol and D-inositol-1,4,5-trisphosphate, and both of these then act as intracellular messengers (second messengers). These lipids can have profound effects at extremely low levels. It can easily be imagined that furocoumarin-lipid adducts could alter the rates of these pathways and consequently have profound effects on regulation of cell processes. Laskin et al. have found that PUVA potently and specifically inhibits binding of epidermal growth factor.[313]

1. Protein Kinase C

Protein kinase C is one of the essential regulatory enzymes that can be affected by this system. The excellent review of protein kinase C by Nishizuka details its crucial role in modulating cell metabolism and its response to various effectors such as phospholipids.[159] A furocoumarin-fatty acid adduct could inhibit a phospholipase and thereby prevent the activation of protein kinase C or other regulatory proteins. Inhibition of protein kinase C could account for the antiproliferative effects of PUVA therapy, as well as inhibition of DNA synthesis; and because of the turnover of proteins and lipids, the temporary nature of the therapeutic benefit of PUVA in treatment of psoriasis might be more easily explained by furocoumarin-fatty acid adducts than by modification of DNA. The effects of PUVA on metabolism of arachidonic acid are different from the effects of UVB or UVC alone,[160] consistent with the unique nature of the formation of cyclobutyl adducts between psoralens and lipids.

2. Cyclic Nucleotides

The level of the regulatory nucleotide cGMP is different in the cells of psoriatic skin than in healthy skin.[161] The activity of guanylate cyclase is ten times higher in involved and three times higher in uninvolved epidermis of psoriatics compared to normal epidermis. Arachidonic acid or HETE stimulates guanylate cyclase activity from involved epidermis two- to threefold and from uninvolved epidermis up to twofold, but these fatty acids have no effect on the activity of this cyclase from normal epidermis. These results indicate that there is an increase in the cGMP biosynthetic capacity of involved epidermis from psoriatics that derives from a markedly increased specific activity of guanylate cyclase and an alteration in a property of this enzyme activity which renders it responsive to fatty acids reported to accumulate in this lesion. These observations are consistent with the report that an elevated steady-state level of cGMP is one of the consequences of the strikingly altered metabolism of cGMP in psoriatic epidermis. This may reflect a defect in the control of one of the signaling pathways that can be temporarily reset by treatment with PUVA and formation of compounds that inhibit activation of cell proliferation. The level of cAMP, another critical regulatory nucleotide, is altered as the result of treatment with psoralens,[162] consistent with this hypothesis.

3. Enzyme Induction

Levine et al. reported that PUVA therapy caused a change in the polypeptide profiles of psoriatic epidermis that suggests significant changes in the regulation of cellular metabolism as occurs with alterations in the state of cell metabolism.[163] These changes occur over a period of weeks and eventually return to their original state. Lowe et al. found that illumination of hairless mice after application of 8-MOP or 5-MOP caused a marked decrease in the incorporation of ^{3}H-thymidine into epidermal disks obtained from mice treated with 8-MOP or 5-MOP,[164] suggesting that DNA synthesis was inhibited. This group also observed an induction of the regulatory enzyme ornithine decarboxylase which is considered a biochemical marker for exposure to cancer promoters. In contrast, similar treatment with the ineffective antipsoriasis agent 3-CPs caused an opposite effect, if any, on ODC induction and incorporation of ^{3}H-thymidine. Several other groups have also observed induction of ODC by illumination after treatment with 5-methylisopsoralen in mice,[165] with 8-MOP or anthracene,[166-168] with 5-MOP,[169] in Chinese hamster cells treated with psoralen plus UVA (or UVC or gamma radiation),[170,171] and in Chinese hamster ovary (CHO) cells treated with TMP.[172] These effects have been interpreted most frequently as an indication of modification of DNA by psoralens, but it seems possible to explain these effects by alterations at other stages of the regulatory process such as points in signaling pathways that might be affected by furocoumarin-lipid adducts. In fact, the changes that are induced by PUVA in the levels of three marker enzymes, acid phosphatase, glucose-6-phosphate dehydrogenase, and capillary alkaline phosphatase, in psoriatic lesions are also caused by application of clobetasol propionate, a corticosteroid, which is not believed to act by chemical modification of DNA, but instead to alter the activity of signaling pathways in the cell.[173] Even if furocoumarins do not act by direct alteration of these messenger pathways via formation of furocoumarin-lipid adducts, it seems reasonable to attempt to design agents that do. Such agents might provide relief of psoriasis symptoms by reducing rates of cell proliferation while avoiding the affects on DNA which may be less specific and more hazardous.

It is also interesting to note that, while pyrimidine cyclobutyl dimers and their bioeffects can be at least partially reversed by treatment with suitable UV light, similar attempts at the reversal of the cytotoxic effects of psoralens have failed in spite of the fact that such irradiation does reverse at least some of the psoralen-DNA adducts.[16] This could be explained by the relatively low efficiency of the reversal reaction or the fact that such reversal is usually performed with light that could itself be toxic due to the formation of pyrimidine cyclobutyl dimers, but it might also be due to the fact that DNA adducts are not responsible for the bioeffects. While furocoumarin-fatty acid adducts are photolysed by UV light, the products of this photolysis do not appear to resemble the reactants,[148] and therefore bioeffects arising from these products should not be photoreversible, in contrast to furocoumarin-DNA adducts.

C. Evidence of Membrane Modifications

While the isolation or detection of covalent adducts between lipids and furocoumarins has not been previously reported, other observations have suggested that such compounds could be formed in cells. For instance, as mentioned previously, Laskin and colleagues observed specific, high-affinity binding sites for psoralens on the surface of HeLa cells and have detected specific binding of 8-MOP to four other human cell lines and five mouse cell lines.[130] This binding was attributed to interaction with either proteins or lipids in the cell membrane. The dark binding of psoralens to membrane receptors could favor the formation of covalent bonds with membrane components via cycloaddition reactions. Upon illumination these binding sites are reported to be covalently modified.

Rytter and associates observed significant inhibition of phytohemagglutinin-induced transformation which temporarily reduces the binding of sheep erythrocytes to lymphocytes, even without exposure to UV light. This binding is known to involve interaction with the cell membranes and can be explained by 8-MOP blocking some lectin binding sites which could be due to modification of membrane proteins or lipids.[174]

Lange et al. observed binding of 8-MOP to the surfaces of lymphocytes and suggested the binding may involve proteins.[175] However, they were unable to establish the mechanism of this binding. Wennersten also obtained results that are consistent with photobinding of psoralens to membranes. He observed that the morphology of cell membranes determined by electron microscopy was changed by high doses of 8-MOP and light.[176] Although doses similar to those used to obtain a therapeutic effect caused no noticeable changes in morphology, it is possible that the same reactions that caused morphological changes when occurring at high levels could have profound physiological effects even at lower doses that do not alter membrane morphology. Wennersten suggested that these effects were not due to singlet oxygen since he did not observe hemolysis of red blood cells (hemolysis is typically observed with photodynamic sensitizers that produce singlet oxygen) or an increase in the morphological changes when water was replaced with deuterium oxide (deuterium oxide lengthens the lifetime of singlet oxygen and therefore can increase the rate of singlet oxygen oxidations).[177,178]

Scherwitz et al. observed psoralen-induced changes in the cytoplasm as well as the cell membrane and wall by electron microscopy of *Candida albicans* cells.[179] They noticed that treatment with PUVA caused the number of vacuoles to increase, and the cytoplasmic membrane disappeared completely in some areas. Meffert et al. observed binding of 8-MOP to cellular membranes and the cytoplasm in the stratum corneum and malpighian layers of guinea pig skin by fluorescence microscopy. Based on these results they suggested that molecules other than DNA might be the targets of 8-MOP and TMP.[133]

Bertaux et al. found that slices of psoriatic human skin incubated with ^{3}H-labeled 8-MOP, irradiated with UVA, and examined by autoradiography exhibited essentially equal binding in the cytoplasm or cytoplasmic membrane and the nucleus.[132] Keratin, collagen, and lipoproteins were labeled, and they claim that these results suggest that proteins as well as nucleic acids might be targets of psoralens. They do not appear to have considered the possibility of covalent binding to lipids, although their results could be explained, at least in part, by this type of reaction.

Danno et al. found that PUVA caused changes in the binding of lectins to the cell membranes of tissue taken by biopsy from guinea pigs treated with topical applications of 8-MOP and irradiation with UVA (1.5 to 3.5 J/cm^2).[180] They concluded that "PUVA treatment perturbs the composition or organization of epidermal cell surface glycoconjugates to induce alterations in lectin stainings." They also point out that "the basic mechanism of PUVA treatment is not fully understood. This is partly because of the complexity of the skin, which consists of a variety of cell populations, structure, and reactions sites. Nuclear DNA certainly is a susceptible target for psoralens. However, psoralen-induced DNA crosslinking and subsequent suppression of DNA synthesis alone do not seem to explain the whole mechanism of action of the photochemotherapy." These investigators had previously found that guinea pig back skin was depleted of epidermal cell membrane fluorescence with pemphigus antibodies and anti-guinea-pig epidermal cell sera by PUVA, and they suggested that these results could be explained by effects on the cell surface domains.[181]

In a study performed at Berkeley by Hearst's group, 4'-hydroxymethyltrioxsalen was found to block cell cycle progression of Cloudman murine melanoma cells.[182] Treatment sufficient to cause modification of only one pyrimidine out of 10^6 DNA bases resulted in the accumulation of cells with DNA content characteristic of the G2 phase. Higher treatment resulting in modification two to three of 10^6 bases caused blockage at the S and G1 phases. The cell cycle blockage was attributed solely to effects on DNA, but it is surprising that such profound effects could be obtained with such small amounts of modification of DNA unless the modifying was specific for the genes that controlled cell cycle progression. It might be worthwhile to determine whether formation of small amounts of lipid adducts could cause an inhibition of cell regulatory pathways sufficient to account for these results instead.

It is interesting to note that the metabolism of triacylglycerol and phosphatidylcholine is much different in psoriatic skin than in normal skin from psoriatic patients or from healthy patients. Coon and colleagues found that much larger amounts of free, long-chain (longer than 18 carbons) fatty acids were present on the surface and at the barrier membrane of psoriatic skin compared to healthy skin.[183-186] In normal skin there is usually only a small amount of long-chain (greater than 18 carbons long) fatty acids on the skin surface, and usually only about one third of the lipids at the barrier membrane consist of these longer-chain fatty acids. However, in psoriatic skin, both the skin surface and the barrier membrane consist of relatively large amounts (up to three fourths) of long-chain fatty acids. This is attributed to an overproduction or to the lack of incorporation of these fatty acids into keratin or glycolipoprotein or both. "In either case the results suggest some defect in lipid metabolism occurring in the psoriatic skin in the region of the barrier zone. It would be premature at this stage of the investigations to speculate on the true nature and cause of this defect."[187]

Summerly et al. found that incorporation of acetate into psoriatic lesions was 50% higher in the adjacent uninvolved epidermis and 120% higher than in epidermis from healthy patients.[188] In psoriatic lesions, a much higher proportion of the total incorporation was into the neutral lipids and was due mainly to a very high incorporation of acetate into the triacylglycerols. Other measurements indicated incorporation into phospholipids, especially phosphatidylcholine, was lower in psoriatic epidermis. They concluded that these observations indicated much higher triacylglycerol synthesis in psoriatic epidermis and a defect in phospholipid metabolism mainly involving phosphatidylcholine and the deacylation (phospholipase A)-reacylation (phospholipid acyltransferase) cycle for fatty acid transfer. In light of these observations one could speculate that this metabolic defect could result in the accumulation of lipid products which keep protein kinase C or other pathways that stimulate cell proliferation in a constant state of activation and PUVA serves to form compounds that can block these pathways. Perhaps the critical metabolic defect in psoriasis is a simple lack of an enzyme that is necessary for the proper processing of fatty acids into lipids.

To carry this speculation further, it is also intriguing to observe, in view of the ability of psoralens to stimulate pigmentation, that lipid droplets (perhaps indicating an accumulation of excess lipids) were observed in a number of patients with hypermelanosis of the skin.[62] Is it possible that psoralen-induced hyperpigmentation could also be due to formation of psoralen-lipid adducts? This would seem to be nearly the opposite of the effect observed in psoriasis therapy where psoralens serve to reduce the rate of cell proliferation, but opposite effects have been observed, depending on the circumstances and subtle changes in structure, for other lipid-like substances such as prostaglandins, depending on the tissue, cell type, and state of the organism.

Since most psoralens are relatively hydrophobic (indeed, their water solubility is usually low), they might be expected to partition into membranes rather well. Association with fatty acids in cell membranes would be expected to enhance the probability of cycloaddition reactions with unsaturated fatty acids. Association with the pyrimidines in DNA certainly plays an important role in determining the rate of cycloaddition of furocoumarins to pyrimidines; the rate of cycloaddition to free pyrimidines in solution is much slower than the rate of reaction in double-stranded DNA.[2] It should be possible to control the relative reaction with DNA and unsaturated fatty acids by altering the ability to noncovalently associate with (intercalate) DNA and the lipids in membranes. By placing bulky, hydrophobic substituents on furocoumarins that prevent efficient intercalation into DNA but do not disrupt solubilization in membranes, it might be possible to minimize the reaction with DNA which may cause mutagenic and carcinogenic effects while maximizing the formation of furocoumarin-lipid adducts which might cause the therapeutic benefit. It will be essential to find substituents that alter dark association while retaining the favorable photochemical reactivity. The selection of appropriate substituents will have to be performed with care since it is often difficult to predict how subtle changes in structure will alter the complex web of steps that are involved in determining the net photochemical reaction and fate in a living organism. One example of how a simple structural change can cause unexpected changes in chemical behavior is 3-CPs. This molecule causes lysis of liposomes even in the dark, in contrast to most other psoralens, possibly due to the delicate balance between solubility in the membrane and destabilization of the hydrophobic forces that are responsible for membrane integrity.[189]

D. Lipid Oxidation

Besides covalent binding to unsaturated fatty acids, oxidation of fatty acids may also occur in some cases, perhaps depending on the conditions. Lipid peroxidation can rapidly lead to degradation of the cellular membrane and consequently death of the cell. Many serious and harmful health consequences including cancer, strokes, atheriosclerosis, inflammation, asthma, arthritis, and the senescence associated with aging have been attributed, at least in part, to lipid peroxidation.[190-192] A wide variety of oxidants, including reactive forms of oxygen such as singlet oxygen, are capable of initiating lipid peroxidation.[193] Since these oxidants can be produced in varying amounts by furocoumarins illuminated with UVA, it is reasonable to suspect that psoralen-induced lipid oxidation could be responsible for the cytotoxicity and perhaps other bioeffects of furocoumarins.

Pathak and Joshi have published several papers that discussed the role of lipid oxidation in furocoumarin bioeffects.[194-197] They suggest that lipid oxidation might be responsible for some of the effects of furocoumarins such as erythema, inflammation, edema, and vesiculation in the skin, while cycloaddition to pyrimidines in DNA might account for cytotoxicity, mutagenicity, and the antiproliferative effects in psoriasis treatment. This hypothesis seems reasonable since lipid oxidation products have been linked to stimulation of the inflammatory response and DNA damage is believed to be the primary cause of mutations and cancer. Cycloaddition of furocoumarins to DNA inhibits DNA synthesis, and therefore it is also reasonable to attribute the antiproliferative effects of psoralens to this reaction. Nevertheless, it should be kept in mind that these are only hypotheses and these mechanisms have not yet been proven. In fact, as mentioned above for the cycloaddition of furocoumarins to unsaturated fatty acids, there are other possible mechanisms that also seem reasonable for the antiproliferative effects of psoralens. A careful examination of the available evidence is warranted to determine what is known with certainty and what is still open to doubt.

Salet et al. found that TMP impaired respiration and uncoupled phosphorylation in isolated rat liver mitochondria much better than psoralen, even though the oxidant yield is higher from psoralen. This result was attributed to the greater solubility of TMP than psoralen in the membrane, which would favor its ability to modify membrane components that are critical to the functions of respiration and oxidative phosphorylation.[198] These results did not allow distinguishing between an oxidative mechanism and cycloaddition mechanism. While it was noted that hematoporphyrin is also capable of disrupting respiration and oxidative phosphorylation, the mechanism for this action is uncertain and could well be different from the mechanism for the actions of psoralens.

Matsuo et al. have found evidence of lipid peroxidation as the result of treatment of human skin with 8-MOP and light.[199] They detected the formation of malondialdehyde, a compound that is believed to be characteristic of the peroxidation of unsaturated fatty acids, and they found that malondialdehyde formation was slightly increased in D_2O compared to H_2O. This solvent isotope effect is too small to establish the intermediacy of singlet oxygen, but it certainly does not rule out the possibility of formation of other oxidants. They also observed inhibition by azide, a well-known singlet oxygen quencher, but it is also well known that azide is not completely specific for inhibition of singlet oxygen reactions, especially in complex biological systems.[200] While these results do not establish the intermediacy of singlet oxygen, they do seem to indicate that psoralens are capable of causing oxidations of fatty acids.

Meffert and colleagues have collected results that they interpreted as indicating production of both singlet oxygen and radicals. They suggested the interesting possibility that formation of radicals led to the binding of furocoumarins to the membrane and the bound furocoumarin then absorbed additional light that led to the production of singlet oxygen which would destroy neighboring membrane components.[201] Potapenko and colleagues also observed furocoumarin-induced lipid oxidation.[202] A later paper by Sukhorukov et al. suggested that this was occurring by a unique and interesting mechanism. They found that photooxidized furocoumarins caused lipid oxidation when added to lipids in the dark, suggesting the formation of a relatively long-lived oxidative intermediate with a lifetime much longer than that of singlet oxygen.[203] Several furocoumarins, including psoralen, angelicin, and 8-MOP, were observed to have this property. A product of 8-MOP illumination in organic solvents was observed to generate chemiluminescence when it was added to water or liposome suspensions.[204] A dioxetane intermediate could behave this way, but so far the structure of the compound responsible for these observations has not been established. Wasserman and Berdahl have reported NMR evidence for the formation of an unstable peroxide intermediate at low temperatures ($-65°C$) which decomposes upon warming, but they did not test for light emission.[205]

It is interesting to note, as Kornhauser has pointed out, that psoralen phototoxicity cannot be inhibited by indomethacin in contrast to the erythema induced by many other photosensitizers that are known to generate oxidants. Prostaglandin formation is usually involved in the inflammatory response to tissue injury. Indomethacin is believed to suppress erythema by inhibiting prostaglandin synthetase, a key enzyme in the pathway for prostaglandin formation in the inflammatory response.[20] This observation therefore suggests that the intermediates that are responsible for psoralen-induced erythema are not the same as the inflammatory oxidants produced by other photosensitizers. Further investigations are warranted to confirm these observations and to identify the intermediates that are responsible for psoralen-induced erythema and inflammation. It is also possible that the same oxidants are produced, but the products of furocoumarin cycloaddition to lipids or fatty acids inhibit one or more of the enzymes that are required for the activation of the prostaglandin pathway.

IV. REACTIONS WITH AMINO ACIDS AND PROTEINS

As discussed earlier in this chapter, several groups have observed intracellular localization of psoralens outside the nucleus, both before and after illumination.[126-128,130-134,206] These observations suggest that furocoumarins bind to cellular components besides DNA, such as lipids and proteins. Cycloadditions to DNA have been thoroughly considered elsewhere, and we have just finished considering binding to lipids, so we will now turn our attention to proteins.

A. Ocular Effects

The most serious biological effect that has been attributed to psoralen binding to proteins is the formation of ocular cataracts. The reaction of 8-MOP with ocular lens proteins has been reported by Lerman's group, and they suggest that this type of reaction could cause cataracts in patients undergoing long-term PUVA therapy. They found that the ocular lenses of patients undergoing PUVA therapy develop a relatively permanent fluorescence which may indicate the covalent binding of 8-MOP to lens tissue.[207-209] Singlet oxygen has been found to cause oxidation of lens crystallins, and the changes that are observed in these proteins are similar to those found in cataractous lenses.[210] Since illuminated psoralens produce singlet oxygen, they may also damage ocular lenses by this type of mechanism. These reactions could be particularly important in long-term PUVA therapy since the ocular lens is completely encapsulated and never sheds its cells throughout life. The rate of turnover of lens proteins is very low. Therefore, following 8-MOP therapy, damaged proteins would be retained and accumulate with repeated therapy.[211]

Lerman's laboratory found that phosphorescence spectra of proteins extracted from the lenses of patients that had undergone PUVA therapy were identical to spectra of isolated lens proteins that had been treated with 8-MOP and UVA light in vitro.[65,122,212] These spectra suggested that 8-MOP had become covalently bound to the proteins. Similar results had previously been obtained in experiments with monkeys.[123] However, the results recorded in rabbit and clinical studies suggest that incidence of cataracts arising solely from PUVA treatment is very low or insignificant.[71-73] The lack of cataract development in rabbits (and perhaps in humans) can be accounted for by differences in species susceptibility — guinea pigs seem to develop PUVA cataracts more readily than rabbits.[69] While the possibility exists that a long latent period delays the observation of elevated cataract incidence, the risk of PUVA cataracts should be easily reduced by providing suitable protection from UVA during the 12 to 24 hr after application of psoralens, required for clearing of psoralens from the eye.[68,69,74,209]

A more thorough evaluation of the hazards and side effects of PUVA treatment would be possible with a complete understanding of the photobiochemistry of furocoumarins and the rates and extents of modification of the various components of biological systems. Lerman's group have observed changes in the fluorescence, phosphorescence, EPR, and NMR spectra of lens alpha, beta, and gamma crystallin proteins and have found similar spectral changes in tryptophan (trp), induced by irradiation with UVA in the presence of 8-MOP.[211,213] They attributed these changes to formation of 8-MOP adducts with trp. However, the changes observed were relatively nonspecific and could be accounted for by simple photolysis of 8-MOP and trp. Controls in which 8-MOP and trp were irradiated alone showed that degradation was slower than in the mixture of these compounds, and therefore some interaction between 8-MOP and trp is probably occurring, but the formation of covalent bonds between 8-MOP and trp was not established since no products were isolated. Oxygen was found to be essential for product formation. In samples that had been degassed *in vacuo* or by purging with inert gas, no spectral changes were observed. Since 8-MOP is known to form singlet oxygen and singlet oxygen readily oxidizes trp, oxidation of trp by singlet oxygen is another possible explanation for these results. It would be helpful to determine the chemical structures of the products formed in this reaction or other evidence to prove that covalent adducts had been formed, and to obtain quantitative kinetic data so that the rates of formation of the products under various conditions could be established. These data might allow more accurate evaluation of the contribution of these reactions to damage in human lenses arising from clinical treatments. A relatively large light dose is needed to cause this reaction: 22- to 40-hr exposures to 360-nm light (20 to 40 hr at 500 mW/cm^2 = 35 to 70 kJ/cm^2 total exposure). It would be useful to determine the quantum yield for this reaction and the fluorescence yield for the product and quantitatively compare this with the fluorescence observed in lens proteins. Since the quantum yield appears to be much higher for covalent binding to unsaturated fatty acids[147,148] than to trp, perhaps the fluorescence observed in lens tissue could be accounted for more easily by the formation of psoralen-lipid adducts.

B. Dark Binding

In addition to ocular proteins, furocoumarin reactions with a number of other proteins have also been studied. However, most of these studies have found that in vitro binding of furocoumarins to proteins and protein inactivation induced by furocoumarins are much slower than binding to DNA.[59,214] Nevertheless, a slower reaction could still be critical in terms of biological effect. It is necessary to inactivate only one critical enzyme in order to profoundly alter biological function. Cyanide, for instance, is a relatively unreactive molecule, and when tested on a number of enzymes it fails to cause much of an effect. However, it is particularly effective at inhibiting a critical enzyme which catalyzes the reoxidation of cytochrome a_3 in the electron transport chain of the respiratory pathway of most aerobic organisms and is therefore a highly toxic compound.

The cycloaddition of furocoumarins to pyrimidines is much faster in DNA than to pyrimidines freely solubilized in water or organic solvents. This is attributed to binding of the furocoumarins to DNA in a geometrical configuration that favors the formation of covalent bonds between the furocoumarin and the pyrimidine.[2,3] Furocoumarins have also been observed to bind to some proteins by noncovalent association in the dark. In fact, dark binding to proteins can be so tight that it interferes with the determination of covalent photobinding.[125] Human and bovine lipoproteins, in particular, have been observed to noncovalently bind furocoumarins with a high affinity.[8,215-220] This noncovalent binding can often be measured using the techniques of gel exclusion chromatography, difference spectrophotometry,[219] equilibrium dialysis of radio-labeled furocoumarins, or fluorescence quenching.[217,218] While the latter techniques are more time consuming than gel exclusion chromatography, they are considered more accurate and reliable. The fluorescence quenching observed has been interpreted as an indication that the trp residue of human serum albumin is involved in the binding.[218] Pulse radiolysis has also been successfully used for measuring binding constants of furocoumarins to both DNA and proteins,[220] and this technique provides an independent means of confirming the binding constants.

Besides favoring the photoaddition of the furocoumarin to the protein, it has been suggested that this binding may play an important role in the delivery of furocoumarins to the target organs in the body.[215] This binding has usually been found to involve one or more specific sites on the protein. For instance, Scatchard analysis of the first step of the binding of 5-MOP to human serum albumin low-density lipoprotein (LDL) yields a binding constant of $1.4 \times 10^5 \ M^{-1}$ and four binding sites.[215] The binding of 5-MOP to human serum LDLs follows a two-step process. Binding is believed to involve the LDL apoprotein. The second step corresponds to a solubilization in the lipidic core of approximately 45 molecules of 5-MOP per LDL molecule. It is accompanied by a large blue shift of the 5-MOP fluorescence. Hydrophobicity can usually be included as an important factor determining the binding affinity.[217] Noncovalent binding to relatively hydrophobic serum albumin is much higher than to the more hydrophilic proteins, thermolysine, ribonuclease, or chymotrypsin;[8] and the extent of binding of various furocoumarins is inversely related to their water solubility. At 1 µg/mℓ and 37°C, 85% of the added angelicin, 84% 8-MOP, 89% psoralen, 91% 5-MOP, 92% 8-methylpsoralen, and 97% TMP were bound to human serum albumin and 88% angelicin, 90% 8-MOP, 94% psoralen, 93% 5-MOP, 94% 8-methylpsoralen, and 98% TMP to human serum.[217] That binding is critically dependent on protein conformation is suggested by the observation that only 2 M guanidinium chloride abolishes binding of 8-MOP even though only small changes in protein conformation were detected.[218]

The ability of LDL to bind 5-MOP and to carry it into various cells may explain the biological effects sometimes encountered during PUVA therapy. In fact, high-lipid diets or some drugs may alter the transport of psoralens via LDL due to the effect on the hydrophobicity of the protein. Delivery of 5-MOP via endocytosis may be involved in lymphocyte activation sometimes observed during PUVA therapy.

Further investigation is needed to establish the extent to which dark association contributes to the photobinding of furocoumarins to proteins. In some cases furocoumarins may even cause alterations of important enzyme activities without the need for light activation.[221] Fouin et al. report that cytochrome P-450 levels both in vivo and in vitro are reduced by 8-MOP with a potency comparable to the powerful P-450 inhibitors SKF 525A and piperonyl butoxide.[222] This result was reflected in the longer hexobarbital sleeping time induced in rats by administration of 8-MOP. The reduction of P-450 activity by 8-MOP was increased by pretreatment with the P-450 inducer phenobarbital and lowered by pretreatment with the

P-450 inhibitor piperonyl butoxide. 8-MOP was observed to covalently bind to microsomal proteins in a reaction that required NADPH and oxygen. Similar to the effects on P-450 activity, pretreatment with phenobarbital increased binding to microsomal proteins, and pretreatment with piperonyl butoxide decreased binding. These observations suggest that P-450 converts 8-MOP into a reactive intermediate that is capable of inactivating P-450 in a suicide-like process. Such effects could have important implications regarding concomitant exposures to 8-MOP and potentially toxic agents that are either detoxified or activated by the P-450 system.

In contrast to the results of Fouin et al., Bickers and colleagues found that over the course of 6 days, administration of 8-MOP caused a two- to threefold increase in the activities of P-450 in rats and mice. Increases in the activities of aryl hydrocarbon hydroxylase and ethylmorphine *N*-demethylase, which were also observed, depended on the species of animal tested.[223] Two other furocoumarins, TMP and isopsoralen, did not alter these enzyme activities. In a separate study, mixed function oxidases were induced in mouse liver by 8-MOP but not by psoralen or TMP.[224] Tsambaos et al. observed an induction of P-450 in rat liver only at relatively high doses (6 to 12 mg/kg daily) with no effect on activity or content in liver microsomes.[225] The differences observed in these studies suggest that the effects depend critically on the dose and time course of administration of furocoumarins as well as the organism and its history of exposure to other agents that alter liver enzyme levels.

C. Photoreactions with Proteins
1. In Vivo

Interpretating the effects of furocoumarins plus UVA on in vivo enzyme levels is difficult, at best, because of the complex set of factors that determine the outcome and the lack of information regarding how these various factors are related. 8-MOP has been observed to cause a decrease in the rate of glucose oxidation in brain homogenate, suggesting that significant damage to enzymes involved in glucose metabolism was induced. A 10% inhibition was observed in the dark, increasing to 43% inhibition after 4 hr of illumination with UVA.[135] Feeding psoralen to albino rats for 1 week caused a 23% reduction in the activity of glucose oxidase in the liver with no decrease in the brain. However, oxygen consumption in the mitochondria of rat liver cells is not affected by doses of psoralen and 365-nm light that are sufficient to cause a large inhibition of DNA synthesis.[226]

It is hard to establish the significance of the effects of PUVA on trp metabolism. An increase in the urinary metabolites of trp (kynurenine pathway) is observed in rats for several days following treatment with PUVA in doses sufficient to cause dermatitis.[227] Metabolite levels returned to normal when skin damage repair was complete. The trp load increased markedly following PUVA treatment, and xanthurenic acid levels were increased the most among the trp metabolites that were measured.[228] The excretion of trp metabolites differs in humans afflicted with sunlight-sensitive skin diseases from healthy individuals, suggesting that this pathway may be involved in the pathology of these diseases.[229] While the implications of this observation are not clear, apparently this is not due to alterations in collagen metabolism.[230] Some of these changes may be involved in the adjustments in metabolism associated with epidermal dysplasia. PUVA has been observed to affect some marker enzymes, such as acid phosphatase, glucose-6-phosphate dehydrogenase, and capillary alkaline phosphatase, in ways similar to the effects of corticosteroids, and in fact the effects of PUVA sometimes last much longer than the effects of the steroids.[173]

2. In Vitro

Several studies of the in vitro effects on proteins have been reported. Furocoumarins have been observed to covalently bind to some proteins, such as lysozyme, serum albumin, histone, and ribonuclease.[124] In addition to covalent binding, furocoumarins may also cause amino acid oxidation. Singlet oxygen is a well-established intermediate formed by irradiation of furocoumarins freely solvated in aqueous solution,[220] and singlet oxygen reacts rapidly with some amino acids, including histidine and trp.[231,232] In addition to singlet oxygen, other oxidative intermediates may also be produced. Veronese et al. found that the relative rates of oxidation of several amino acids by 8-MOP was different than the relative rates of photooxidation sensitized by the well-known photochemical sources of singlet oxygen, rose bengal or methylene blue.[233] This observation is a good indication that, at least in some reactions, oxidative intermediates besides singlet oxygen are produced by photoexcited furocoumarins. Since covalent binding and oxidation would not necessarily involve the same type of amino acids, which mechanism accounts for enzyme inactivation should depend on the nature of the amino acids that are critical for enzyme activity and the relative rates of reaction of these amino acids via each of the mechanisms. Careful design of experiments is required to provide conclusive tests of the mechanisms responsible for furocoumarin-photosensitized enzyme inactivation.

a. Bovine Serum Albumin (BSA)

Covalent binding of furocoumarins to proteins sometimes involves formation of metastable furocoumarin intermediates. Irradiation of BSA with UVA in the presence of 8-MOP results in the formation of a covalent adduct via a relatively long-lived reactive intermediate that is formed by the photolysis of the furocoumarin. Preirradiation of 8-MOP followed by mixing with the protein in the dark results in covalent binding of the 8-MOP to the protein; preirradiation of the protein in the presence of typical singlet oxygen sensitizers followed by mixing with the furocoumarin does not.[234,235]

Acetylation of the tyrosine hydroxyl in BSA decreased the photobinding of 8-MOP, suggesting that tyrosine was one of the binding sites. At least one other amino acid is also involved in the binding, however, since there is only one tyrosine residue in BSA and more than 1 mol of 8-MOP was bound per mole of protein. The product formed in the binding reaction was fluorescent and had a UV absorbance maximum at 310 nm, similar to 4′,5′ adducts of furocoumarins with pyrimidines. However, formation of the furocoumarin-BSA adduct must be different from that of the pyrimidine adduct since binding was reduced in the absence of oxygen, in contrast to formation of the pyrimidine adduct.

Veronese et al. studied the covalent photobinding of some linear and angular furocoumarins to a selection of proteins, including BSA, glutamate dehydrogenase, enolase, thermolysin, and ribonuclease.[236] An earlier study by this group with BSA, thermolysin, ribonuclease, and chymotrypsin had shown that the affinity of noncovalent association in the dark was not correlated with the rate of formation of covalent adducts upon illumination.[214] This indicates that factors other than the strength of noncovalent association limit the rate of photobinding, but it does not rule out the possibility that dark association is a necessary factor for photoaddition. The determination of the role of dark association in photobinding would be easier if the rates of noncovalent association and disassociation in the dark were known. These rate constants could be compared with the rate of photoadduct formation to determine, first of all, whether the rate of dark association was truly slower than photobinding. The equilibrium constant for noncovalent association in the dark is not, by itself, sufficient to conclusively establish that this process is not required for photobinding, since a rapid rate of dark association could provide the opportunity for the few molecules that are bound to undergo the slower photobinding reaction.

The rate of furocoumarin binding to proteins was correlated with the rates of photolysis of the furocoumarins when the furocoumarins were irradiated in solution without the proteins. This observation, in addition to the finding that binding in the absence of oxygen was slower than in aerated solution (but still significant), led them to conclude that photobinding of furocoumarins to proteins proceeded by two mechanisms: (1) involving direct binding of the photoexcited furocoumarin to the protein; and (2) involving intermediate formation of a reactive compound via photolysis of the furocoumarin which, in turn, binds to the protein. They suggested that the second mechanism involved the formation of singlet oxygen which then oxidized the furocoumarin to a reactive agent. The identification of the chemical structure of the furocoumarin-amino acid adducts would assist in the analysis of the mechanism of this reaction, but so far none of these structures has yet been determined.

b. Subunit Cross-Linking

Schiavon and Veronese measured the extent of subunit cross-linking in three enzymes — glutamate dehydrogenase, catalase, and alcohol dehydrogenase — induced by 5,8-dimethoxypsoralen, 8-MOP, 8-methylpsoralen, psoralen, and 3-CPs.[237] The extent of subunit cross-linking was correlated with the ability of the furocoumarins to photosensitize formation of singlet oxygen as determined by Vedaldi et al.[238] However, these data are based on the *p*-nitrosodimethylaniline (RNO) method for singlet oxygen measurement, which is not necessarily specific for singlet oxygen. These results might therefore be considered representative of oxidant production rather than exclusively singlet oxygen. The rate of glutamate dehydrogenase cross-linking induced by psoralen was increased in D_2O compared to H_2O, but the increase was much smaller than that expected for the increased lifetime of singlet oxygen. This is not proof that singlet oxygen is not involved, however, because the magnitude of the rate enhancement is a combination of the effects of the solvent, not only on the lifetime of singlet oxygen, but also on sensitizer triplet quantum yield and the SΔ ratio, as well as the rate constant for reaction of singlet oxygen with the substrate. Because of the complex dependence of each of these factors on the solvent, the D_2O test for singlet oxygen is ambiguous. It would be useful to know whether the subunit cross-linking observed by Schiavon and Veronese could also arise from treatment with pure sources of singlet oxygen, the nature of the products of the reaction and the chemical steps that lead to formation of these products. Analysis of the chemical structures of the cross-links would be a first step toward providing this information.

Cross-linking of enzyme subunits does not necessarily destroy enzyme activity. Schiavon and Veronese noted that the rate of subunit cross-linking was faster than the rate of inactivation for catalase and cross-linking occurred more slowly than inactivation for alcohol dehydrogenase. This indicates that subunit cross-linking is not an inactivating event for alcohol dehydrogenase unless there is more than one type of inactivating event. Nor is cross-linking an inactivating event for catalase unless there is more than one type of cross-linking reaction. For glutamate dehydrogenase, no conclusion can be drawn regarding the role of subunit cross-linking in enzyme deactivation, since they both occur at the same rate. One conclusion can be drawn regarding the mechanism of subunit cross-linking in these enzymes: cross-linking does not require bifunctional reaction of the furocoumarin with the protein, since the number of furocoumarin molecules covalently bound was insufficient to account for the extent of cross-linking. In addition, oxygen appears to be necessary for cross-link formation since little cross-linking occurred under nitrogen. The role that oxygen plays in this process, is however, uncertain. The cross-linking is not apparently due to disulfide bridge formation since the reducing conditions employed in the dissociation analysis used for determining the extent of cross-linking would have severed such bonds.

c. DNA Polymerase

Granger et al. examined the inactivation of the three enzymatic activities of *Escherichia coli* DNA polymerase I by UVA-illuminated 8-MOP.[239] They found that 8-MOP plus UVA caused a reduction in the 5'-3' polymerase and 3'-5' exonuclease activities. The 5'-3' exonuclease activity was first stimulated slightly, then inactivated as illumination continued. No inactivation was observed when 8-MOP was added to the protein in the dark after various periods of UVA preillumination of the 8-MOP, nor did illumination of a mixture of the enzyme and photolysed 8-MOP reduce enzyme activities. This shows that 8-MOP photoproducts do not cause inactivation of the enzyme either in the light or the dark and that the inactivation observed with illuminated 8-MOP must arise from 8-MOP primary photochemistry since polymerase absorption of the light used in this experiment was too weak to account for the observations.

CD (circular dichroism) spectra of the modified and unmodified protein show that inactivation is not due to a conformational change in the protein and therefore the alteration in enzyme activity must be due to modification of amino acids that are essential to enzyme function, most likely at the active site or allosteric control sites. Inactivation of all three enzymatic activities requires oxygen. The rate of inactivation increased more than tenfold in D_2O compared to H_2O. A concentration of 0.01 to 0.1 M diazabicyclooctane (DABCO), an efficient singlet oxygen quencher, prevented inactivation of 5′-3′ exonuclease, and partially reduced the rate of inactivation of the polymerase, and 3′-5′-exonuclease. A concentration of 1 mM sodium azide had no effect on 5′-3′ polymerase activity, and 10 mM azide caused some inhibition of inactivation. The effect of azide on the other enzyme activities was not measured. 8-MOP photoproducts did not cause inactivation in the light or in the dark. The 8-MOP triplet must not be reacting directly with polymerase since the reaction is much faster in the presence of oxygen, which should be quenching the triplet. The inactivation is most likely due to amino acid oxidation. Inactivation of the 5′-3′ exonuclease activity is different than inactivation of the polymerase or 3′-5′ exonuclease activities. The complete protection of the 5′-3′ exonuclease activity by DABCO and sodium azide as well as the dependence on oxygen suggest that inactivation is due to reaction with singlet oxygen. Since DABCO and azide were only able to partially protect the other enzyme activities, the inactivation of these activities cannot be solely attributed to reaction with singlet oxygen. Note, however, that the 8-MOP triplet lifetime is increased by a factor of 2 in D_2O.[240] The greater than tenfold increase in the inactivation rate of the polymerase activity by D_2O is therefore ambiguous and does not necessarily indicate that the effects on polymerase activity are due solely to reaction with singlet oxygen, either. Inactivation may arise from 8-MOP bound near the critical amino acids.

The stimulation of the 5′-3′ exonuclease activity observed in the early stage of the reaction with 8-MOP may be due to release of the enzyme from the 3′-OH of the primer as the result of damage to the 3′-5′ exonuclease active site. This might provide more enzyme for the 5′-3′ exonuclease reaction. Alternatively, it may be due to consecutive modifications of amino acids at the 5′-3′ active site, the first modification being responsible for stimulation and subsequent modifications responsible for inactivation. Granger et al. point out that damage to the accuracy of the polymerase could be far more effective at causing mutations than a number of damages to DNA bases which are efficiently repaired. Mutations are believed to arise due to alterations in the genetic information contained in DNA. These alterations can arise due to errors in the repair or synthesis of DNA. Errors can occur due to chemical modification of DNA bases which alter their base pairing ability. The mutations that arise from treatment with furocoumarins plus UV light have been most frequently attributed to chemical modification of DNA by cycloaddition to the pyrimidine bases as discussed earlier in this chapter. However, modification of the accuracy of the enzymes that repair or synthesize DNA such that they insert the incorrect bases into new DNA should also cause mutations. It would be interesting to know whether the chemical modification of DNA polymerase caused by 8-MOP as reported by Granger et al. occurs at a sufficient rate to account for mutations arising in cells. However, the rate of inhibition of DNA template activity under the same conditions used to inactivate the polymerase was similar to the polymerase inactivation rate, and therefore the reaction of 8-MOP with *E. coli* DNA polymerase I appears to occur at a rate that is comparable to the rate of direct chemical modification of DNA. Whether the effects on DNA polymerase or on any other enzymes can actually account for any of the biological effects of PUVA will require further investigation.

Granger and Helene later reported a further dissection of the mechanisms of the effects of 8-MOP on DNA polymerase.[241] They found that irradiation of DNA polymerase I in the presence of 8-MOP forms a covalent adduct between 8-MOP and polymerase. A maximum of 3 mol of 8-MOP was bound to each mole of polymerase. It is unlikely that noncovalent binding is required for formation of the covalent adduct since the dark binding equilibrium constant is less than $10^4 \ M^{-1}$. At least two different adducts appear to be formed on the basis of fluorescence spectra of the product mixture. Adduct formation is accompanied by inactivation of enzyme activity when irradiated under oxygen. When this reaction is performed in the absence of oxygen, the rate of adduct formation is the same, but the enzyme is not inactivated. Further irradiation of the adduct formed anaerobically (which still had the same enzymatic activity as unmodified enzyme), after removal of free 8-MOP and addition of oxygen, results in inactivation of the polymerase activity at a faster rate than that obtained with free 8-MOP. The 5'-3' exonuclease activity was not affected in this experiment. The inactivation of the polymerase activity was not inhibited by DABCO (concentration was not specified) and did not require oxygen, although oxygen increased the rate of inactivation. Therefore, this inactivation is attributed to two different reactions: (1) oxidation of amino acids by an intermediate reactive oxygen species which is not free singlet oxygen and (2) an oxygen-independent reaction possibly involving a radical intermediate. Reaction of singlet oxygen with amino acids in the near vicinity of the bound 8-MOP via a caged intermediate cannot be ruled out since such a reaction might not be inhibited by DABCO.

d. Lysozyme

Similar to the results obtained with DNA polymerase, Poppe and Grossweiner found that 8-MOP-photosensitized inactivation of lysozyme was much slower in the absence of oxygen than in its presence.[250] They established that the inactivation proceeded via singlet oxygen by showing that the rate constant for reaction of the reactive intermediate with lysozyme was the same within experimental error as that for reaction of singlet oxygen with lysozyme (1.3×10^9 ℓ/mol/sec). They found that addition of poly(dA-dT) slowed lysozyme inactivation apparently due to inhibition of singlet oxygen formation. That poly(dA-dT) was simply quenching singlet oxygen was ruled out by noting that eosin-sensitized inactivation of lysozyme, which had previously been established as a singlet-oxygen-mediated reaction, was not significantly inhibited by poly(dA-dT). This report by Poppe and Grossweiner is one of the most convincing demonstrations of the intermediacy of singlet oxygen in a furocoumarin-photosensitized reaction with biological relevance and is one of the most cited papers on this topic.

In contrast with the results obtained with 8-MOP, Muller-Runkel and Grossweiner surprisingly found that 3-CPs caused inactivation of lysozyme faster when oxygen was excluded.[189] The fact that the rate of inactivation increased as the concentration of the enzyme decreased indicates that inactivation was due to formation of an inactivating agent in the external medium rather than direct reaction of the photoexcited 3-CPs with the protein. Muller-Runkel and Grossweiner suggested that the inactivation proceeded through formation of a 3-CPs radical rather than by direct reaction of the 3-CPs excited triplet with lysozyme, since they claimed that the triplet reacts too rapidly with oxygen to account for the significant rate of inactivation under aerobic conditions. Certainly singlet oxygen is not responsible for lysozyme inactivation in the absence of oxygen, but Muller-Runkel and Grossweiner's results do not rule out singlet oxygen as an intermediate under aerobic conditions. Under aerobic conditions, oxygen would quench most of the 3-CPs triplet producing singlet oxygen, as Muller-Runkel and Grossweiner suggest. Since the triplet would no longer be available for producing radicals or interacting directly with the protein, all of the lysozyme inactivation in the presence of oxygen could be due to singlet oxygen. The results of Poppe and Grossweiner establish the ability of singlet oxygen to inactivate lysozyme. Since 3-CPs has been shown to be a potent source of singlet oxygen in oxygenated solution,[251] it is certain that at least some of the inactivation occurring in the presence of oxygen is due to singlet oxygen. In the absence of oxygen, the 3-CPs triplet would be free to undergo other reactions which could lead to radical formation. The faster rate of inactivation in the absence of oxygen could be explained if the 3-CPs radical was more potent as a lysozyme inactivator than singlet oxygen. It is important to realize that the mechanism of the reaction can depend on the conditions, and especially on the oxygen concentrations. A more accurate determination of the role of singlet oxygen in the 3-CPs-sensitized inactivation of lysozyme could be performed by measuring the rate of lysozyme inactivation by pure singlet oxygen and the yield of singlet oxygen formation from illuminated 3-CPs under the conditions reported in the paper. Comparison of these data to the rate of lysozyme inactivation observed with 3-CPs would allow the determination of whether singlet oxygen was responsible for lysozyme inactivation in this reaction.

e. Ribosomes

The results of Singh and Vadsaz indicate that singlet oxygen may play a major (but not exclusive) role in the inactivation of ribosomes by illuminated 8-MOP and TMP.[252,253] The rate of ribosome inactivation was reduced 75% under nitrogen compared to oxygen. A concentration of 0.01 M azide, an excellent singlet oxygen quencher, provided similar protection. A plot of inhibition of ribosome inactivation vs. azide concentration revealed a biphasic relationship: the inhibition plateaued at about 80% inhibition. The biphasic relationship suggests that the slight inhibition observed at higher concentrations probably involves a process not involving singlet oxygen. The ability of azide to quench other reactive intermediates besides singlet oxygen, such as sensitizer triplet states, has been noted by others.[200,254] The singlet-oxygen-quenching amino acids, methionine, histidine, and trp, also inhibited inactivation. A large rate enhancement was observed in D_2O, consistent with the large increase in the lifetime of singlet oxygen in this solvent. These results were taken as an indication that approximately 75% of the ribosome inactivation arises from reactions with singlet oxygen.

Superoxide dismutase (SOD) did not provide protection, leading to the conclusion that superoxide is not involved in the psoralen-induced inactivation. Poly(dAT) provided about 20 to 30% inhibition of inactivation, presumably by competition for free psoralen, but rRNA and poly-U did not. Therefore, 8-MOP does not appear to intercalate into rRNA. Ethanol and isopropanol, efficient hydroxyl radical scavengers,[255] provided no protection.

Inactivation of tRNA synthetase as a model for ribosomal proteins was very similar, in response to inhibitors, to inactivation of ribosomes. This contrasts with the inactivation of tRNA as a model of rRNA, which was not affected by inhibitors in the same way as ribosome inactivation. In fact, the results suggested that tRNA inactivation involved endogenous photosensitizers; tRNA was inactivated faster without 8-MOP than with it. 8-MOP may protect by absorbing some of the light; it apparently does not cause much inactivation of rRNA on its own. These results suggest that the ribosomal inactivation is due primarily to reactions with proteins rather than rRNA.

f. Amino Acid Oxidation

Veronese et al. measured the photosensitized inactivation of several enzymes by 8-MOP, rose bengal, and methylene blue.[233] Of the eight enzymes studied, lysozyme was the most rapidly inactivated. For 8-MOP, of the 17 amino acids examined, only cystine, methionine, and histidine were rapidly oxidized, while tryptophan and tyrosine were not significantly degraded. This is in contrast to the pattern seen with rose bengal and methylene blue, which caused the most rapid oxidation of histidine and tryptophan, most likely via singlet oxygen. These data strongly suggest that other oxidants besides singlet oxygen were responsible for the 8-MOP-induced amino acid oxidation. It is not known whether the degradation of the amino acids measured in these experiments was actually responsible for the inactivation of the enzymes. It is possible that multiple reactions are occurring, and only one of these may actually lead to loss of enzyme activity.

A later study by Veronese and colleagues[236] produced somewhat different results, this time, more in agreement with the results expected from reaction with singlet oxygen as the sole oxidant: tryptophan was oxidized along with histidine, tyrosine, methionine, and phenylalanine, and the same pattern of amino acid oxidation was observed with lysozyme and bovine liver glutamate dehydrogenase. Methionine was converted to methionine sulfoxide. Methionine sulfoxide is a product of the singlet oxygen oxidation of methionine. They found that only 0.8 mol of psoralen was bound to lysozyme, 0.9 mol to glutamate dehydrogenase, and 0.6 mol to ribonuclease, although amino acid analysis showed the loss of 5, 7, and 5 amino acids, respectively. Electrophoresis analysis of the trypsin-hydrolyzed lysozyme after irradiation in the presence of ^{3}H-labeled psoralen showed several radioactive spots, suggesting that there may be several psoralen binding sites in the protein. The oxidation of free amino acids was also measured. These results showed that tryptophan, histidine, and methionine were oxidized, with phenylalanine reacting somewhat more slowly and oxidation of other amino acids not detected, again consistent with intermediacy of singlet oxygen. The inactivation of glutamate dehydrogenase was enhanced fourfold in D_2O compared to H_2O and was inhibited by the singlet oxygen quenchers, sodium azide, and histidine. Only psoralen-sensitized inactivation was inhibited by the hydroxyl radical scavenger sodium benzoate. Mannitol, the other hydroxyl scavenger, did not have an effect on any of the reactions, nor did the enzymes SOD, peroxidase, or catalase. The antioxidant ascorbic acid was by far the best inhibitor. The concentration of azide gave 50% protection at the same concentration that provided this much protection in a pure singlet oxygen system.[256] While some inactivation occurred in a sample purged of oxygen with nitrogen, this could be attributed to traces of oxygen that were not removed or that entered the system during the removal of aliquots for assays. Since glutamate dehydrogenase is easily deactivated by hydrogen peroxide, it is clear that hydrogen peroxide was not generated in these reactions. Glutamate dehydrogenase is not sensitive to superoxide,[257] however, and it is therefore not possible to evaluate its production from these results. Both the extent of photobinding to the proteins and the ability to modify free amino acids was correlated with rate of enzyme inactivation. This is a good example of a case in which two different chemical reactions are correlated with a bioeffect, due either to accidental coincidence or to the fact that all three reactions involve the same rate-limiting step at some early stage.

Results similar to those obtained by Veronese et al. regarding the effect of inhibitors were obtained by Schiavon et al. in a study of the inactivation of ribonuclease A by psoralen.[258] This inactivation reaction was inhibited most by ascorbate, reduced substantially under nitrogen compared to oxygen, and also inhibited by the singlet oxygen quencher azide. An amount of 1.4 mol of psoralen was bound to each mole of protein at a dose causing 40% inactivation. Based on identical CD spectra of the modified and unmodified enzyme, the inactivation was apparently not due to a conformational change in the protein. Instead, inactivation must be due to chemical modification of amino acid(s) that are essential to enzyme function. Amino acid analysis of 60% inactivated protein indicated loss of 1 methionine, 1 tyrosine, 1 lysine, 1.5 histidine, 0.5 glutamic acid, 0.5 half-cystine, and 0.4 phenylalanine, results consistent with the oxidation of the protein by singlet oxygen; those amino acids most susceptible to oxidation by singlet oxygen were lost most extensively by treatment with psoralen and UVA. However, this result is not sufficient to prove singlet oxygen was responsible for amino acid modification or inactivation of the enzyme. The results obtained with HPLC and electorphoresis analysis of peptides obtained by tryptic digestion of the protein after treatment with ^{3}H-labeled psoralen and illumination were inconclusive regarding the location of the covalent biding of psoralen because appropriate controls were not included to determine the retention time or electrophoresis pattern of psoralen photolysis products. The peptide analysis indicates which peptides are most extensively modified, but it was not carried further to determine which amino acids in each peptide were altered to account for the modification. Therefore, the damage cannot be localized to specific amino acids. Overall, the results are rather ambiguous regarding the mechanism of inactivation of ribonuclease by irradiated psoralen. This study shows that some amino acids are modified and some psoralen is bound to the protein, but, again, the structures of the products and the details of the chemical mechanisms responsible for these results are not known.

g. Dopa Oxidation

Besides the reactions with nucleic acids, lipids, and proteins, photomodification of some other cell constituents has also been reported. Craw et al. have measured some of the photophysical consequences of the illumination of 8-MOP in the presence of L-3,4-dihydroxyphenylalanine (L-dopa).[260] They find that the 8-MOP triplet state was quenched at the diffusion-controlled rate by 3,4-dihydroxyphenylalanine, producing the 8-MOP radical anion and the L-dopa radical cation. They point out that the dopa radical cation is a precursor of melanin,[261] and therefore this reaction might account for the increased pigmentation induced by psoralens. Additional studies are required to evaluate this possibility by determining the extent of this type of process in human skin.

h. Flavin Mononucleotide

Modification of flavin mononucleotide (FMN) was one of the first photoreactions of furocoumarins that was reported.[262-265] This reaction results in the formation of a relatively unstable adduct, but since its discovery in the middle 1960s this reaction has received little attention, perhaps because the cycloaddition of furocoumarins to DNA was believed to have greater potential for exerting biological effects.

i. Furocoumarin Photolysis and Covalent Binding

The nature of the reactive intermediate(s) formed in furocoumarin-sensitized reactions responsible for covalent binding to proteins has not yet been identified, but some clues have been obtained. Wasserman and Berdahl[205] and Logani's group[266] isolated a photoproduct, 6-formyl-7-hydroxy-8-methoxycoumarin, in 11 and 1.4% yield from the irradiation of 8-MOP in aerated deuterochloroform-deuteromethanol and dichloromethane, respectively. The same compound was previously isolated in the 8-MOP-sensitized photoreaction of FMN.[263] This product may arise from singlet oxygen oxidation of 8-MOP since formation of this compound was inhibited by the singlet oxygen quencher DABCO. However, in contrast to the observations of Yoshikawa et al., this and other relatively stable photoproducts obtained from photolysis in dichloromethane were not found to react with tyrosine or tryptophan in methanol or water. It is possible that the reaction between 8-MOP photoproducts and amino acids observed by Yoshikawa et al.[234] involves a less stable intermediate such as the dioxetane, peroxide, or endoperoxide of 8-MOP, which serves as a precursor of 6-formyl-7-hydroxy-8-MOP or one of the other stable photoproducts. The unstable intermediate may be formed via 8-MOP oxidation by singlet oxygen; the binding reaction is faster in 90% D_2O than in H_2O, consistent with intermediacy of singlet oxygen in the formation of the reactive furocoumarin intermediate. However, the rate increase did not approach the theoretical value for a true singlet oxygen reaction; in principal the rate should have increased by a factor of at least 15 due to the at least 15 times longer lifetime of singlet oxygen in D$_2$O compared to H$_2$O.[177] The structure is consistent with attack of singlet oxygen on the 4'-5'-bond of 8-MOP to give a dioxetane followed by decomposition to the 6-formyl-7-hydroxy derivative.

Wasserman and Berdahl also observed NMR evidence in a chloroform-methanol solution of 8-MOP irradiated at low temperature ($-65°$C) consistent with the formation of a peroxidic compound (positive starch-iodide test) but were unable to assign the structure of this compound. This compound rapidly decomposed upon warming, apparently yielding only a polymer which could not be characterized. Further investigation is necessary to determine whether this intermediate can account for the covalent binding of 8-MOP to proteins and the chemiluminescence observed by other investigators in photoreactions of 8-MOP with lymphocytes[175] or upon addition of photolysed organic solutions of 8-MOP to aqueous samples.[204] In the meantime, the mechanism of formation of covalent adducts between 8-MOP and amino acids is uncertain.

j. Conclusions

Looking at all of the experimental data together does not allow many firm conclusions to be drawn regarding the chemical mechanisms or even the biological significance of furocoumarin photoreactions with proteins. In few cases has singlet oxygen been ruled out as a potential intermediate, and most observations have been rather ambiguous. Identification of the chemical structures of both the furocoumarin and protein products formed in these reactions would be helpful, but additional information would still be required for a complete understanding of this chemistry. In addition, quantitative kinetics data are needed under a variety of conditions to evaluate the possibilities of these reactions under biological conditions: effects of ionic strength and solvents on reaction rates will allow determination of the differences arising from localization of furocoumarins at various sites within cells. The relative rates of modification of the various targets within cells, including nucleic acids, proteins, and lipids, and the quantitative information regarding the biological consequences of these modifications would help to establish which types of reactions are responsible for the biological effects.

IV. SINGLET OXYGEN GENERATION

Two fundamental chemical processes are frequently mentioned in discussions of the mechanisms of furocoumarin-protein reactions: formation of singlet oxygen and formation of radicals. Both of these fundamental types of processes may play an important role in the chemical modification of lipids, as well, and these processes will therefore be discussed next to determine whether this information can shed any light on the mechanisms of the furocoumarin-induced modification of biological molecules.

A. Quantum Yield

In 1971 Song et al. reported the results of analysis of luminescence spectroscopy and molecular orbital calculations of the excited states of coumarin, psoralen, and 4-hydroxycoumarin.[138] Among their conclusions and predictions they suggested that the inhibition of the photomutagenic effect of 8-MOP by oxygen observed in bacteria was due to quenching of the 8-MOP triplet by oxygen to give singlet oxygen. This was one of the first discussions of the possibility of production of singlet oxygen by photoexcited furocoumarins. Oginsky et al. had previously observed lower rates of bacterial killing by furocoumarins under oxygen than under nitrogen.[154] Bevilacqua and Bordin found that oxygen (and paramagnetic ions) inhibited the photocycloaddition of psoralen to pyrimidine bases, leading them to conclude that the photocycloaddition reaction proceeded through the triplet state.[140] Triplet quenching by molecular oxygen is well known to form singlet excited oxygen, and therefore these observations strongly suggested that singlet oxygen production was possible with photoexcited furocoumarins.

Poppe and Grossweiner were the first to provide convincing evidence that psoralens could produce singlet oxygen in significant amounts.[250] They observed a transient by flash photolysis with a lifetime of 1.8 μsec in deaerated water and 0.4 μsec in oxygen-saturated water which they assigned on the basis of corroborative data to the 8-MOP-excited triplet. The shorter lifetime in the presence of oxygen is consistent with quenching of the triplet by oxygen, most likely yielding singlet oxygen. They also found in continuous irradiation experiments that 8-MOP illuminated with UVA formed an intermediate that inactivated lysozyme, but only in the presence of oxygen; illumination under nitrogen gave no significant inactivation. The rate constant for the reaction of this intermediate with lysozyme was shown to be in good agreement with the value previously determined for the rate constant of lysozyme inactivation by singlet oxygen generated by eosin photosensitization[267] and by direct laser excitation of the triplet-singlet transition of molecular oxygen.[268]

The generation of singlet oxygen by furocoumarins was subsequently measured by several groups by a variety of methods. Cannistraro and Van de Vorst estimated the singlet oxygen yield from a set of furocoumarins by measuring the formation of nitroxide radical from 2,2,6,6-tetramethylpiperidine in ethanol using ESR spectroscopy.[269] They found the relative yields of nitroxide radical were in the order 5-MOP > psoralen > 8-MOP > angelicin. The yield of nitroxide radical from 5-hydroxy- and 8-hydroxypsoralen was not significant. Unfortunately, the results were not corrected for differences in absorption of the polychromatic light used for excitation, and therefore these results cannot be directly compared with other measurements. In addition, the formation of nitroxide radicals could arise from reactions not involving singlet oxygen, and therefore this assay is not particularly reliable. A later study by DeCuyper et al. used the more reliable detection of 3-β-hydroxy-5-α-cholest-6-ene-5-hydroperoxide (the reduced form of the 5-hydroperoxy derivative of cholesterol) to monitor singlet oxygen production as described by Foote et al.[270] Again, the results apparently were not corrected for differences in absorption of light or concentrations of furocoumarins, and the results differ significantly from those previously reported, in particular for the high yield of singlet oxygen reported from 5-hydroxypsoralen. This contradicts a number of other measurements of the singlet oxygen formation and triplet yield for 5-hydroxypsoralen, including experimental and theoretical studies, and no satisfactory explanation is available for this discrepancy.

De Mol and Beijersbergen van Henegouwen obtained some results that convincingly confirmed the earlier report by Poppe and Grossweiner of the formation of singlet oxygen from illumination of 8-MOP.[271] They detected the formation of the characteristic products of singlet oxygen oxidation of 2-methyl-2-pentene[272] in the 8-MOP-sensitized reaction by gas chromatography and mass spectrometry analysis and comparison to authentic compounds. They found that formation of these compounds was inhibited by β-carotene, and none was formed in the absence of oxygen.

They also observed 8-MOP-induced photooxidation of dopa. Singlet oxygen was implicated as an intermediate in the dopa oxidation by agreement of the β value determined from kinetics of the 8-MOP-sensitized reaction with the β value determined in the methylene blue-sensitized reaction. This agreement may be fortuitous, however, since it is possible that the methylene blue-sensitized oxidation of dopa does not involve exclusively singlet oxygen as the reactive intermediate. There is evidence that the reaction of singlet oxygen with the neutral tyrosine, which resembles dopa, is very slow, but that excited dye reacts directly with it very rapidly in a Type I mechanism.[272-276] A similar mechanism should be possible for the 8-MOP-sensitized oxidation of dopa, and in fact the quenching of 8-MOP triplet by dopa has been reported to give semioxidized dopa and the 8-MOP radical anion.[260]

Preliminary results of a pulse radiolysis analysis show that the semioxidized dopa radical undergoes ring closure to ultimately form dopachrome via dopaquinone. De Mol and Beijersbergen observed the weak pink color that is characteristic of dopachrome after irradiation of their mixtures of 8-MOP and dopa, results consistent with this mechanism.

If the furocoumarin oxidation of dopa is not a specific measure of singlet oxygen formation, the subsequent reports by De Mol and Beijersbergen of singlet oxygen yields from illuminated furocoumarins are not reliable.[277] The agreement of β values for furocoumarin-sensitized dopa oxidation with the β value of the methylene blue-sensitized reaction is not sufficient proof of the specificity of this assay for singlet oxygen, since, as mentioned before, methylene blue may also cause Type I oxidations. The relatively small solvent isotope effect observed (2.5) for the reaction in D_2O is consistent with a nonsinglet oxygen mechanism for this reaction; based only on the difference in the singlet oxygen lifetime in D_2O and H_2O, and neglecting other complications, such as effects on sensitizer excited state lifetimes, rates of intersystem crossing and quenching by oxygen, the rate of the oxidation reaction should be approximately 15 times faster in D_2O.[177] The order of singlet oxygen yield determined by De Mol and Beijersbergen — TMP > psoralen > 8-MOP > 5-MOP > 5,8-dimethlpsoralen > angelicin — differs somewhat from the later, more reliable determinations performed by Knox et al.[251] and Beaumont et al.[144] These differences may arise from the lack of specificity of the dopa oxidation assay.

Pathak and Joshi used two different assays to compare the singlet oxygen yield from a number of different photosensitizers, including furocoumarins.[194-197] One of these assays employs the spectrophotometric determination of the bleaching of p-nitrosodimethylaniline (commonly abbreviated as RNO to symbolize the nitroso function attached to an organic molecule) due to oxidation by an intermediate formed in the reaction of an imidazole, such as histidine, with singlet oxygen.[278] This assay is subject to some of the same uncertainties associated with the measurement of dopa oxidation. Although the oxidative intermediate responsible for RNO bleaching is believed to only form by reaction of singlet oxygen with the imidazole, it seems likely that other oxidants could accomplish the oxidation of RNO as well. It might be possible to improve the specificity of the assay by measuring the half-life of the oxidative intermediate responsible for RNO bleaching or by characterizing it in some other way, but this is usually not performed.

The other assay employed by Pathak and Joshi was the measurement of oxidation of deoxyguanosine (dGuo).[279] While dGuo oxidation is certainly not specific for singlet oxygen,[151] they suggest that the rate of this reaction may reflect the general ability of furocoumarins to cause oxidative DNA damage. It should be noted that some of Pathak and Joshi's results were not corrected for differences in absorption or scattering of light (the samples containing crude coal tar were turbid and therefore scattered the incident light considerably) or differences in concentrations, and therefore the results cannot be quantitatively compared to results from other studies. However, the determinations may be more suited to comparisons with in vivo data, since in vivo experiments are usually performed with light sources similar to the ones used by Pathak and Joshi and with application of similar quantities of sensitizers.

The formation of superoxide was determined by spectrophotometrically measuring the reduction of nitroblue tetrazolium.[280] Nitroblue tetrazolium can be reduced by a number of other agents besides superoxide, including ascorbate, and it is therefore necessary to provide additional evidence that the reduction is due specifically to superoxide. Pathak and Joshi found that the reduction reaction was inhibited 76 to 100% by the superoxide-specific enzyme superoxide dismutase (SOD) depending on the sensitizer. Those reactions that were not completely inhibited by SOD may involve other reductants besides superoxide.

The results of the two different assays for oxidant (singlet oxygen) production were similar with respect to the relative magnitude of the effectiveness of the sensitizers and gave the following order of oxidant yields: hematoporphyrin > phenanthridine > acridine > methylene blue > CCT > fluoranthrene > anthracene > anthrone > pyrene > 8-MOP > anthralin > chloroquine > anthralin dimer.[194] The order for the furocoumarins tested was 3-CPs > hematoporphyrin > psoralen > angelicin > 5-methylangelicin > 4,8-dimethyl-5'-carboxypsoralen > 3-amino-4'-methylpsoralen > anthracene > TMP > 4,5'-dimethylangelicin > 8-MOP > 3,4'-dimethyl-8-MOP > 5-MOP > 5'-diethylaminobutoxypsoralen. It is not clear how the furocoumarin results fit to the other sensitizers since absolute values were not given, and the furocoumarin results were reported as a separate set of data from most of the other sensitizers. The latter results for furocoumarins were corrected for differences in concentrations, and the differences in absorbance over the wavelength range of the incident light were reported, but corrections for variations in light intensity over this wavelength range were not applied, and therefore these results also cannot be quantitatively compared with other studies. The rates of oxidations in both assays were enhanced by D_2O, although the rate enhancement (~40 to 70%) was far lower than that expected for the increased lifetime of singlet oxygen (~15 times), and reactions were much slower when oxygen was excluded.

That singlet oxygen is not the only reactive oxygen intermediate generated by these sensitizers was confirmed by the observation that superoxide yields were significant and the production of superoxide tended to parallel results in the singlet oxygen assays, although there were distinct exceptions. For instance, 3-CPs, which was an excellent singlet oxygen sensitizer, gave only small amounts of superoxide, while anthracene was an exceptionally strong source of superoxide. It is interesting to note that the nitroblue tetrazolium reduction was inhibited 70 to 95% by 0.025 *M* DABCO but not by 0.10 *M* azide, confirming the suspicion that DABCO is not necessarily a specific singlet oxygen quencher.

These results were also compared to the ability of these sensitizers to cross-link DNA. Cross-linking was determined from the rate of elution of DNA from a hydroxyapatite column at elevated temperature after heat denaturation and rapid cooling. This assay was only able to provide qualitative results compared to the other assays, but within the limitations of the data, these results suggest that the order of the ability to cross-link DNA differed considerably from the ability to produce oxidants, such as singlet oxygen, or redox-active agents such as superoxide, with psoralens being the most potent DNA cross-linkers of the agents tested.

The most direct determination of singlet oxygen yields reported to date are those of Knox et al.[251] They used the recently developed solid state detection of the near infrared luminescence of singlet oxygen[281-284] to obtain quantitative values for singlet oxygen quantum yields in benzene. In addition they determined the triplet-minus-singlet extinction coefficients which allowed calculation of the rates of intersystem crossing in deoxygenated solution for several furocoumarins. S_Δ, the fraction of triplet quenching that produces singlet oxygen, was determined by comparison of triplet quantum yields and singlet oxygen quantum yields. S_Δ was found to vary with the structure of the furocoumarin, a not unexpected result in consideration of previous observations.[285] This variation in S_Δ must be considered in attempting to determine the singlet oxygen yields based solely on triplet yields or in attempts to establish mechanisms based on correlations of triplet yields with biological functions. Generation of singlet oxygen was observed exclusively via the triplet state; generation via the singlet state was ruled out by lack of dependence of the singlet oxygen quantum yield on oxygen concentrations at concentrations sufficiently high to allow significant singlet quenching by oxygen. The order of singlet oxygen quantum yields was 3-carbethoxypseudopsoralen > 3-CPs > TMP > pseudopsoralen ~ 5-MOP > psoralen > 8-MOP ~ 5,8-dimethoxypsoralen. S_Δ values for these compounds ranged from 1 for 3-CPs and 0.96 for 3-carbethoxypseudopsoralen to 0.1 for 5,8-dimethoxypsoralen with the other furocoumarins having values between 0.3 to 0.6. The value of 0.17 determined from bleaching of diphenylisobenzofuran by Ronfard-Haret et al. for the singlet oxygen quantum yield of 3-CPs in benzene[286] is about 50% lower than the value determined by Knox et al. The reason for this disagreement is not apparent, since if other nonsinglet oxygen mechanisms were contributing to diphenylisobenzofuran bleaching, the value for singlet oxygen quantum yield should have been overestimated rather than underestimated.

In contrast with the results of Pathak and Joshi, Knox et al. found the quantum yield of superoxide production for psoralen and 8-MOP to be below the detection limit, i.e., less than 0.001. The discrepancy may be due, at least in part, to the difference in the solvents; the measurements of Knox et al. were performed in benzene, while Pathak and Joshi's determinations of superoxide yield were performed in water but may also be due to the lack of specificity of the superoxide assays used by Pathak and Joshi.

B. Singlet Oxygen Formation by Furocoumarins Bound to DNA

Sherman and Grossweiner estimated the number of encounters of the surface of DNA with singlet oxygen molecules relative to the frequency of occurrence of cycloadducts using large-target diffusion theory.[287] They conclude that the diffusive reactions of singlet oxygen generated by free furocoumarin can compete with photoadduct formation. Therefore, if singlet oxygen production is sufficiently high, it is possible that damage to DNA from reaction with singlet oxygen could compete with the cycloaddition reaction.

The significance of singlet oxygen damage to DNA, therefore, depends on the amount of singlet oxygen produced by furocoumarins in the presence of DNA. Poppe and Grossweiner observed inhibition of the 8-MOP-sensitized photooxidation of iodide by added poly(dA-dT).[250] Since iodide oxidation was attributed to intermediacy of singlet oxygen, this observation suggests that singlet oxygen yield is lowered by DNA, perhaps because 8-MOP bound to DNA produces less singlet oxygen. The possibility that the DNA was simply quenching singlet oxygen was ruled out by finding that poly(dA-dT) did not inhibit the eosin-photosensitized oxidation of iodide which also occurs via singlet oxygen.[250] Poppe and Grossweiner suggested that thymine in DNA was efficiently quenching the 8-MOP triplet and thereby preventing the formation of singlet oxygen by energy transfer similar to the quenching of the eosin triplet by binding to human serum albumin.

Consistent with this suggestion, the flash photolysis results of Goyal and Grossweiner showed that oxygen quenching of the 8-MOP triplet state was reduced when in the presence of an excess of DNA compared to when it was free in solution.[288] This was so even though glycerol was added to the solution lacking DNA, so that viscosity and rates of oxygen diffusion would be approximately the same. This suggestion was disputed by De Mol and Beijersbergen,[289] who found that dopa oxidation was faster in the presence of DNA than in its absence. De Mol and Beijersbergen point out that Goyal and Grossweiner did not calculate the extinction coefficients for the triplets, and since the absorbance spectra differ in the presence and absence of DNA, it is not valid to draw conclusions regarding the extent of triplet quenching. However, the alteration of the triplet extinction coefficients by binding to DNA would have to be much larger than would be expected from such interactions to account for the magnitude of the changes in the intensity of the triplet absorbances evident in Goyal and Grossweiner's spectra. Furthermore, the faster dopa oxidation seen by De Mol and Beijersbergen in the presence of DNA can be explained by the lack of specificity of the dopa assay used by De Mol and Beijersbergen for singlet oxygen, rather than an indication that singlet oxygen is still being produced. Dopa oxidation can be sensitized by 8-MOP via formation of radicals which may be formed even in the presence of DNA without singlet oxygen formation.[260] Other considerations also suggest that De Mol and Beijersbergen's results may be in error. De Mol and Beijersbergen observed dopa oxidation, although slower, even in the absence of furocoumarins, apparently due to a DNA-sensitized reaction (the DNA absorbed about 10% of the incident light) or to contaminants in the DNA. The actual rate of oxidation attributed to DNA-bound 5-MOP involved correcting for the rates of oxidation due to DNA alone and free 5-MOP determined by a complex calculation. Any errors in the calculation or the determination of the DNA- and free 5-MOP-induced reactions would cause an error in the determination of the amount of the reaction to be attributed to bound 5-MOP. These results are therefore subject to some uncertainty. It would have been easier to interpret the De Mol and Beijersbergen's results if conditions had been adjusted so that the DNA alone did not contribute significantly to dopa disappearance, e.g., longer wavelength light or purified DNA, and if the dependence of dopa oxidation rate on DNA concentration was then determined so that results could have been simply extrapolated to infinite DNA concentration.

Recent data obtained by Beaumont et al. support the contentions of Grossweiner and his colleagues for the case of AMT and psoralen.[144] The triplet state yields in illuminated aqueous solutions of AMT and psoralen were measured by flash photolysis and the near infrared luminescence of singlet oxygen was monitored using a germanium diode as a direct and unambiguous measure of singlet oxygen concentration. These results indicate that addition of DNA to a nitrogen-saturated solution of psoralen or AMT causes a reduction in the triplet-triplet absorption. Since singlet oxygen yield was observed to be linearly dependent on the triplet yield, it is apparent that the lower triplet-triplet absorption is not due to a reduced triplet extinction coefficient. In fact, triplet formation from DNA-bound psoralen of AMT was below the detection limit.[144] This would be the case if the triplet lifetime is much shorter for intercalated furocoumarins than for free furocoumarins, if the rate of intersystem crossing is reduced by association with nucleic acids, or if the singlet state rapidly relaxes due to radiative or nonradiative processes such as physical or chemical quenching before it has a chance to undergo intersystem crossing to the triplet. In surprising contrast, no decrease was observed in furocoumarin triplet or singlet oxygen yields from 8-MOP upon addition of DNA, but the yields were an order of magnitude lower than for AMT or psoralen, and no conclusions can be drawn at this time from these preliminary data. Still, this leaves open the possibility that some psoralens may be able to generate singlet oxygen when complexed to DNA, although almost certainly at a reduced level compared to free psoralens.

The difference in triplet yields when bound to DNA, observed for 8-MOP compared to psoralen and AMT, may arise from differences in the rate of quenching of the singlet states by DNA. Salet et al.[290] by fluorescence spectroscopy and Beaumont et al.[144] by time-resolved laser flash measurements observed a decrease in the singlet lifetime for AMT when bound to DNA. Since the triplet state was not detected for DNA-bound AMT, it was suggested that cycloaddition must occur via the singlet state for AMT, and at least part of the quenching of the AMT singlet by DNA should therefore involve the primary interactions that result in cycloadduct formation. If this is so, the smaller yield of AMT triplet and the consequent smaller yield of singlet oxygen observed upon addition of DNA can be attributed to the effective singlet quenching, and the smaller effect of DNA on the 8-MOP triplet might reflect the lower rate of cycloaddition of 8-MOP to DNA compared to AMT.[2,3] That cycloaddition to DNA occurs via the singlet state rather than the triplet is consistent with the product distribution that is observed; cycloaddition of psoralens to DNA is found to form preferentially the 4',5' monoadduct rather than the 3,4 monoadduct. This is in contrast to reactions of psoralens free in solution, which preferentially form the 3,4 monoadduct. Reaction at the 4'5' bond is predicted to occur preferentially from the singlet state, while reaction at the 3,4 bond is predicted to occur preferentially from the triplet state.[137,138,291,292] Therefore, the difference in the geometry of the adducts formed form DNA-bound and free psoralens may represent differences in the excited states responsible for reaction. Reactions via the singlet state are not likely for psoralens free in solution because the singlet state lifetime is so short ($2.0 \pm - 0.2$ nsec)[293] that the singlet state psoralen is not likely to encounter a suitable substrate by diffusion before it decays. Indeed, available evidence is consistent with cycloadduct formation via the triplet in free solution.[140] However, when bound to DNA, the substrate (the pyrimidine base) is held in close proximity, so that reaction via the singlet state is possible.

VI. RADICAL FORMATION

Besides quenching by oxygen, furocoumarin-excited states can also be quenched by amino acids and other cellular constituents. Quenching of furocoumarin triplets by amino acids has been observed to form radicals. One of the earliest observations of radical formation by irradiation of furocoumarins was reported by Pathak et al.[294,295] This group observed radical formation by ESR spectroscopic analysis of irradiated samples of 8-MOP and 4',5'-dihydro-5-ethylcarbamyl-8-methoxypsoralen irradiated with a 500-W medium pressure mercury arc lamp in water, ethanol, glycerol, benzene, and mixtures of these solvents at low temperature. Because of the high intensity of this lamp and the relatively long-lived triplet states at the low temperatures used in these studies, biphotonic processes are probably responsible for the radicals observed. Since biphotonic transitions are not believed to be important under the conditions most often employed in clinical applications and in most studies of biological effects,[286] we will not consider studies[166,296] involving biphotonic radical production further.

Normally irradiation of furocoumarins with low-intensity UVA radiation in neat water or organic solvents does not generate radicals; however, addition of suitable electron donors such as trp, dopa, and even thymine can result in the relatively efficient formation of the furocoumarin radical anion and the substrate radical cation by electron transfer.[260,297-299] The spectra of the radical anion and cations of 8-MOP, 5-MOP, and 3-CPs have been well characterized by flash spectroscopy and pulse radiolysis.[298,300] Although 3-CPs triplet is rapidly quenched by oxygen (3-CPs is one of the best singlet oxygen sensitizers among the furocoumarins that have been studied), the rates of quenching of the 3-CPs triplet by thymine or uracil are much slower (less than 1.0×10^6 below the detection limit) than the rate of quenching of psoralen (thymine, 7.5×10^8; uracil, 1.1×10^8) or angelicin (thymine, 1.1×10^9; uracil, 2×10^8). However, tryptophan and tyrosine quenched the 3-CPs triplet rapidly with rate constants of 3.6 and 2.3×10^9, respectively. As for other furocoumarins, quenching by tryptophan and tyrosine led to formation of the 3-CPs radical anion.[286,301]

While formation of free radicals from furocoumarins does not occur by monophotonic mechanisms at significant yields in water, spectral characteristics of the triplet state have been interpreted in terms of formation of a species that somewhat resembles the radical anion. It has been suggested that this observation indicates exciplex charge transfer from water to the furocoumarin triplet. This conclusion is supported by the results of theoretical calculations on the model compound divinylbenzene.

The details of the steps that may occur subsequent to radical formation have not been completely established. The role that radical formation may play has nevertheless been considered and at least in one case, radical formation has been proposed as an explanation for one of the beneficial bioeffects of furocoumarins, i.e., pigment formation.[260] Craw et al. suggested that the efficient quenching of 8-MOP triplet by L-dopa to give the 8-MOP radical anion and semioxidized dopa results in rearrangement of the dopa and further oxidation to dopachrome, an intermediate on the path to melanin synthesis. Such a reaction might explain in part the special ability of furocoumarins to stimulate pigmentation, although the kinetics of this reaction might not be able to account for the long latent period for pigment development that is usually observed with furocoumarin-induced tanning.

VII. CORRELATIONS OF BIOEFFECTS WITH CHEMICAL REACTIVITY

One of the most popular ways to investigate the mechanisms of the bioeffects of furocoumarins has been by testing correlations of the chemical reactivity of various furocoumarins with specific bioactivities. Bioactivity depends on the type and intensity of the light source, the route of administration, the biological endpoint measured, and the species tested. Because of the complex way in which these factors interplay, one might expect that discovering correlations of biological effects with a single aspect of chemical reactivity would be purely fortuitous and not particularly meaningful. Absorption, metabolism, distribution, penetration to the target, and excretion all play important roles in determining the net biological effect, yet all of these processes involve dark (nonphotochemical) reactions which are likely to be unrelated to the photochemical reactions responsible for modification of the cellular targets. Note for instance that with topical application, TMP is approximately 25 times more effective than 5-MOP and 7 times more effective than 8-MOP at causing skin erythema, but with oral administration TMP is found to be only 2 times more effective than 5-MOP and slightly less effective than 8-MOP.[277] These results were obtained in the same type of organism, presumably with the same type and intensity of light source. Even greater differences might be observed if these other factors are varied.

Various bioeffects have been compared with formation of DNA monoadducts and DNA cross-linking, cycloaddition to pyrimidines, production of singlet oxygen, and triplet yield. Whether or not the data are correlated depends on how the chemical reactivity and bioeffects are measured. Even if such a correlation could be shown not to be due to coincidence (the conditions used to measure the chemical reactivity are usually much different from the conditions in the living tissue with respect to solvent, ionic strength, ability of the chromophore to absorb light, wavelengths and intensities of light used, and concentrations of various substrates) the correlation could still be trivial since correlations are not necessarily unique. Any reaction which proceeds through the same reactive intermediate will also be correlated. For instance, if the cycloaddition to DNA was actually the event responsible for the biological effect, and both cycloaddition and singlet oxygen production proceeded through the triplet excited state, both cycloaddition and singlet oxygen production could be correlated with the biological activity. Any change in psoralen structure will change the triplet yield not only for cycloaddition but also for singlet oxygen production. Unless the reactivity of the triplet toward cycloaddition and singlet oxygen production differs with psoralen structure, both will exhibit the same correlation. The same holds for any other psoralen photochemical reaction that proceeds through the triplet.

Nevertheless, as long as these limitations are kept in mind, correlations of rates of in vitro reactions with bioeffects may be worth considering even though such comparisons are not likely to provide a conclusive answer. In some situations, correlations can be used to rule out some of the mechanisms which would otherwise be considered possible. Correlations have been frequently tested and discussed, almost certainly because of the difficulty in finding any better way to gain some insight into the complex chemistry characteristic of living organisms and their responses to exogenous agents. We will therefore review some of these studies in order to gain any insight they might provide.

Joshi and Pathak compared the rates of oxidant production determined by monitoring RNO bleaching and dGuo oxidation and the ability to cross-link DNA, as discussed earlier in this chapter, with the ability of several furocoumarins to induce skin erythema.[194-197] The results indicated that the correlation of none of the chemical activities perfectly matched the bioactivites, but in spite of the poor correlation, Pathak and Joshi suggested that singlet oxygen or oxidant production may be responsible for skin erythema and inflammation, and vascular damage, effects which are easily accounted for by lipid peroxidation; they also suggested that oxidative reactions might contribute to the mutagenic and carcinogenic effects of some furocoumarins.

While Pathak and Joshi failed to find a good correlation of oxidant production and bioeffects, De Mol and Beijersbergen claimed that the yield of singlet oxygen from various psoralens correlates reasonably well with the effectiveness as skin photosensitizers.[277] However, the methods used to measure singlet oxygen yield in many of these studies, i.e., dopa or guanine oxidation, are not specific for singlet oxygen and therefore could be detecting other types of oxidative intermediates in addition to singlet oxygen. Other more reliable measurements of singlet oxygen production have been reported recently, which indicate that the singlet oxygen yield of furocoumarins in benzene are not correlated with photosensitizing activity in skin.[251] For instance, singlet oxygen production by 3-CPs is several times greater than by 8-MOP, but 8-MOP is a much more effective sensitizer of erythema in skin. The differences in the conclusions reached by De Mol and Beijersbergen and Knox et al. are due, in part, to the furocoumarins that each group used in assessing the degree of correlation; De Mol and Beijersbergen did not include 3-CPs, the most anomalous derivative, in their analysis. Differences may also be due in part to the difference in solvents and the different assays used by these two groups in measuring singlet oxygen production.

De Mol and Beijersbergen also suggested that some of the photochemical effects of furocoumarins on DNA could be due to singlet oxygen production and consequent oxidation of DNA bases.[89,289] However, the most reliable measurements to date indicate that the yield of singlet oxygen from illuminated psoralens is extremely low when bound to DNA. In addition, the rate of reaction of singlet oxygen with nucleic acid bases is much slower than the rate of reaction with amino acids such as histidine and tryptophan, and other potential cellular targets. Therefore, the cycloaddition to pyrimidines should exceed any damage that could arise from singlet oxygen-induced base oxidation for most furocoumarins, unless the amount of free psoralen in the vicinity of the DNA was considerably higher than the amount of psoralen intercalated in DNA.[144] Mutagenicity that has been observed from photodynamic dyes and attributed to singlet oxygen[89] may well involve other types of reactive intermediates, instead. A recent report indicates that rose bengal, a potent singlet oxygen sensitizer, is capable of efficiently causing strand breaks in DNA. This reaction is inhibited by oxygen, proving that strand breaks are formed faster in this reaction via a nonsinglet oxygen mechanism than by the singlet oxygen pathway.[302] Nearly all of the photosensitizers which have been observed to cause bacterial mutations have also been observed to cause nonsinglet oxygen photochemical reactions.

It has also been suggested that the stimulation of pigmentation arising from treatment with furocoumarins may be due to singlet oxygen reactions or other types of oxidations, specifically those involving oxidation of dopa and other melanin precursors.[277] The rate-limiting step in pigmentation is at the level of oxidation of tyrosine and dopa,[303] and besides enzymatic processes, nonenzymatic processes apparently play a role in melanin production. It is also possible that oxidation could destroy endogenous agents, such as thiol-containing compounds that serve as inhibitors of pigment formation; these compounds can be readily oxidized by singlet oxygen. However, if this is the case, barring the complications discussed in the introduction to this section, one might expect furocoumarins with more potent capacities to induce oxidation to be more potent tanning agents, but this is not the case. In spite of a much higher photooxidizing activity of 3-CPs, for instance, it is essentially inactive as a tanning agent compared to other weaker photooxidizers, such as 8-MOP.[304]

Several studies have even shown that the extent of triplet formation in several furocoumarin derivatives is not correlated with skin photosensitizing potency.[299,301] The triplet quantum yield of 3-CPs is eight times larger than the triplet quantum yield for 8-MOP; yet 8-MOP is a considerably more effective photosensitizer. This can have several explanations: (1) the reactivity of the triplet with biochemical substrates limits the effects rather than efficiency of triplet formation; (2) some photochemical intermediate other than the triplet, such as the excited singlet or radical anion, is responsible for the chemical reactions that cause the effects attributed to furocoumarins and (3) factors other than photochemistry determine the effects, such as distribution or metabolism of the furocoumarin within the organism. Further research is required to determine which of these explanations is correct.

Vedaldi et al. compared the ability of 12 furocoumarins to produce singlet oxygen and DNA cross-links in aqueous solution to their ability to inhibit DNA synthesis in cell culture and their ability to induce skin erythema.[238] The results were not adjusted for differences in light absorption by the different furocoumarins tested, and therefore quantitative comparison with other results is difficult. Presumably, similar light sources were used in the chemical and biological experiments. No correlation was observed between singlet oxygen yield and either of the biological effects. One of the most notable examples was 3-CPs, which was the best source of singlet oxygen but exhibited the lowest ability to inhibit DNA synthesis and was totally inactive at inducing skin erythema in the tests reported in this paper. On the other hand, an excellent linear correlation was observed between the ID_{50} for inhibition of DNA synthesis and the reciprocal of the rate constant for cycloaddition to DNA determined in previous studies. It is surprising that such a good correlation can be obtained, considering the complex relationship of all of the factors that determine bioactivity. While there are certainly limitations to these data, this report represents one of the most convincing demonstrations of the importance of the cycloaddition reaction in determining the ability of furocoumarins to inhibit DNA synthesis. However, this correlation does not indicate whether cycloaddition is also responsible for the other bioeffects of furocoumarins. It is worth noting that these results shed little light on the mechanisms of other types of bioeffects such as tanning and the antiproliferative effect obtained in the treatment of psoriasis. Inhibition of DNA synthesis, while certainly one means of inhibiting cell proliferation, may not be the only way. If an agent could be found that inhibited cell proliferation without chemically modifying DNA, such an agent might prove to be a much more effective treatment for hyperproliferative diseases such as psoriasis. Since cycloaddition to pyrimidines and cycloaddition to unsaturated fatty acids probably depend on the same excited state intermediates, it would be interesting to see if cycloaddition to unsaturated fatty acids also correlates with these data.

A. Protection by Antioxidants

While the correlation reported by Vedaldi et al. suggests an important role for cycloaddition in determining the effectiveness of furocoumarins at inhibiting DNA synthesis, the observation that antioxidants provide protection against some of the phototoxicity of furocoumarins suggests that oxidations may also play a role in furocoumarin photocytoxicity. For instance, β-carotene, a potent singlet oxygen quencher,[305] was found to provide protection against phototoxicity, measured as erythema, of orally ingested (in the diet) 8-MOP in albino female Osborne Mendel rats.[306] The protective effect was proportional to the concentration of carotene found in the skin by HPLC. This is not conclusive evidence, however, of a causal role for singlet oxygen or other oxidants in 8-MOP actions. Although this observation is consistent with the conclusion that carotene is protecting from the phototoxicity of 8-MOP by quenching singlet oxygen, it is also possible that carotene protects by altering the structure or metabolism of the cells rather than by direct effects on the damaging agents or that it quenches other reactive intermediates such as the 8-MOP excited states. Indeed, the data reported by Potapenko et al. indicate that 8-MOP-induced photooxidation of alpha-tocopherol proceeds via a nonsinglet oxygen pathway.[307,308] This conclusion is reached by noting that the quantum yield for tocopherol oxidation exceeds the maximum possible quantum yield for singlet oxygen.[301] Potapenko et al therefore propose that tocopherol inhibition of PUVA photoxicity may also arise from tocopherol's ability to trap free radicals.

B. Oxygen Dependence

The oxygen dependence of some bioeffects of furocoumarins may also be taken as an indication of the relative importance of photocycloaddition and photooxidation reactions of furocoumarins. The greater bactericidal activity of 8-MOP in the absence of oxygen, reported by Oginsky et al., suggests that bacterial killing is due primarily to photocycloaddition (to nucleic acids, unsaturated fatty acids, or other suitable substrates) or other reactions not involving oxygen.[16,59,153-155] On the other hand, not all furocoumarins exhibit the same oxygen dependence. The lethal effect of 3-CPs in yeast suggests that photooxidation may be involved for this photoxensitizer, since this effect is reduced by removing oxygen.[286] This result probably reflects the fact that 3-CPs is a potent source of singlet oxygen in aerobic solution and it is relatively slow to form nucleic acid adducts.

The oxygen dependence by itself, however, is not necessarily a simple and straightforward indication of the role of photooxidation. It is important to keep in mind that, while the absence of a change in rate of killing upon removal of oxygen indicates that singlet oxygen formation is not required for killing, it is not proof that in the presence of oxygen an oxygen-dependent mechanism is not involved. It is possible for the chemical mechanism to change depending on conditions. Oxygen can rapidly quench the furocoumarin triplet state.[138,286,288,304,309] This quenching reaction often produces singlet oxygen in high yield,[144,251] and it decreases the chance that the furocoumarin triplet can take part in any other reactions. In fact, Bevilacqua and Bordin found that oxygen slowed the cycloaddition of psoralen to pyrimidine bases by a factor of about 2.[140] In the presence of oxygen, therefore, singlet oxygen production is favored and cycloaddition is less likely. In the strict absence of oxygen, on the other hand, singlet oxygen cannot be formed and the cycloaddition reaction is un-inhibited by oxygen quenching. If the furocoumarin triplet and singlet oxygen are both capable of killing the bacterium with equal efficiency, then no change in the rate of bacterial killing might be observed upon removal of oxygen. Other similar possibilities exist for any other toxic intermediates that could also be formed from the photoexcited furocoumarin. If killing were only observed in the presence of oxygen, it would suggest that killing was not due to the cycloaddition reaction (although interpretation of such results in complex biological systems is far from straightforward). If killing were only observed in the absence of oxygen, it would indicate that singlet oxygen was almost certainly not involved (since singlet oxygen cannot be not formed in the absence of oxygen, this result is more convincing). However, if killing is observed under both aerobic and strictly anaerobic conditions, it is difficult to draw any firm conclusions regarding mechanism. Unfortunately, this seems to be the case for most observations of the oxygen dependence of furocoumarin bioeffects.

These considerations also suggest that tests of correlations of singlet oxygen yields with bioeffects may not be appropriate over a wide range of singlet oxygen yields, since it is possible for the mechanism to differ depending on the type of photochemistry most appropriate for a given compound. This is apparent in the difference in the oxygen dependence of the lethal effect of 3-CPs and 8-MOP in microorganisms mentioned above. Whereas oxygen enhances the lethal effect of the potent singlet oxygen generator 3-CPs in yeast,[310,311] it inhibits the lethal effect of the weak singlet oxygen generator but potent DNA cross-linker 8-MOP in bacteria[16,144,153,311] or phage.[312] It is possible to explain these results by a difference in the type of photoreaction responsible for the cytotoxicity of these two compounds.

VIII. UNANSWERED QUESTIONS: FUTURE RESEARCH

Looking over the research results reviewed in this chapter it appears safe to say that some furocoumarins undergo relatively rapid photocycloaddition to nucleic acids and unsaturated fatty acids and relatively slow covalent binding to proteins. The details of some aspects of the reaction with nucleic acids are well known, but many of the details of the reactions with proteins, and even fewer with lipids, have been established. Further investigations are needed to determine whether furocoumarin-fatty acid adducts are formed in cells and what role such adducts might play in altering regulation of cell proliferation and metabolism. The identity of the reactive intermediates responsible for covalent photobinding of furocoumarins to proteins is still unknown, and in fact the identity of this intermediate may depend on the type of protein involved, since some photobinding reactions require oxygen and others prefer its absence.

Furocoumarins have been conclusively shown to generate singlet oxygen, but the significance of this reaction in terms of the effects on living organisms is uncertain. In some cases, singlet oxygen production may be responsible for inactivation of proteins or oxidation of fatty acids, but it is unlikely to cause significant chemical modification of nucleic acids because of their slow rate of oxidation by singlet oxygen and the lowered ability of most furocoumarins to produce singlet oxygen when bound to DNA. Singlet oxygen yields for some furocoumarins have been established in benzene, but it will be important to establish the solvent dependence of singlet oxygen generation and to determine how it is affected by the presence of other cellular constituents. Still uncertain is the relative importance of each of these classes of reactions in determining the bioeffects.

Determining the chemical reactions that are responsible for the effects of agents such as furocoumarins that have so many possibilities is definitely a considerable challenge. Since simple correlations are relatively uninformative, it would be helpful to have agents that could be demonstrated to undergo only one type of chemical reaction. A compound that was certain to only undergo cycloaddition to nucleic acids or lipids, with no production of radicals, singlet oxygen, or other reactive oxygen species, or one that only produced singlet oxygen would be especially useful. Each of these agents could then be tested for their biological effects separately. Once the characteristic kinetics of the biological effects for each individual type of reaction, such as cycloaddition to DNA, cycloaddition to lipids, and oxidations of amino acids, have been measured, the rates of each of these chemical reactions for the various furocoumarins should be measured in cells. The rate of each chemical reaction in the cell could be compared to the effectiveness of each of these reactions in causing any observed biological effects to determine the contribution of each reaction to the outcome.

Unfortunately, it is particularly rare to find photochemical agents that undergo a single type of chemical reaction. Nearly all photosensitizers that produce singlet oxygen are capable, at least in principle, of undergoing a number of other types of reactions, such as direct attack of the sensitizier-excited states on substrates or formation of radicals, especially in the complex mixtures of compounds characteristic of living systems. Lacking specific agents, it would be useful to be able to measure accurately the rates of each of the different type of reactions as well as the rates of the onset of the biological effects for a set of agents that differ in the relative rates of each of these processes. It is critical that the techniques that are used to measure the rates of each of the chemical reactions be absolutely specific and free of ambiguity. A deconvolution analysis could then establish the connection between the chemical and biological events. Unfortunately, assays that are absolutely specific for detecting such short-lived intermediates as singlet oxygen have been extremely elusive in the past. Consequently, it is easier to describe such a goal than to achieve it. It would be necessary to perform the analysis with a great deal of care and attention to detail, but the data that would be obtained should provide better insight regarding the connection between molecular events and biology than is currently available. The development of methods for the accurate and unambiguous determination of the rates of individual chemical processes should therefore be a high-priority item on the research agenda for furocoumarins.

REFERENCES

1. **Vigny, P., Gaboriau, F., Voituriez, L., and Cadet, J.,** Chemical structure of psoralen-nucleic acid photoadducts, *Biochimie,* 67, 317, 1985.
2. **Hearst, J. E.,** Psoralen photochemistry, *Annu. Rev. Biophys. Bioeng.,* 10, 69, 1981.
3. **Cimino, G. D., Gamper, H. B., Isaacs, S. T., and Hearst, J. E.,** Psoralens as photoactive probes of nucleic acid structure and function, organic chemistry, photochemistry, and biochemistry, *Annu. Rev. Biochem.,* 54, 1151, 1985.

4. **Hearst, J. E., Issacs, S. T., Kanne, D., Rapoport, H., and Straub, K.,** The reaction of the psoralens with deoxyribonucleic acid, *Q. Rev. Biophys.*, 171, 1, 1984.
5. **Hearst, J. E.,** Psoralen photochemistry and nucleic acid structure, *J. Invest. Dermatol.*, 771, 39, 1981.
6. **Song, P. S. and Tapley, K. J., Jr.,** Photochemistry and photobiology of psoralens, *Photochem. Photobiol.*, 296, 1177, 1979.
7. **Scott, B. R., Pathak, M. A., and Mohn, G. R.,** Molecular and genetic basis of furocoumarin reactions, *Mutat. Res.*, 39, 29, 1976.
8. **Rodighiero, G. and Dall'Acqua, F.,** In vitro photoreactions of selected psoralens and methylangelicins with DNA, RNA, and proteins, *Natl. Cancer Inst. Monogr.*, 66, 31, 1984.
9. **Pathak, M. A.,** Mechanisms of psoralen photosensitization reactions, *Natl. Cancer Inst. Monogr.*, 66, 41, 1984.
10. **Pathak, M. A.,** Molecular aspects of drug photosensitivity with special emphasis on psoralen photosensitization reaction, *J. Natl. Cancer Inst.*, 691, 163, 1982.
11. **Grossweiner, L. I.,** Mechanisms of photosensitization by furocoumarins, *Natl. Cancer Inst. Monogr.*, 66, 47, 1984.
12. **Parsons, B. J.,** Psoralen photochemistry, *Photochem. Photobiol.*, 326, 813, 1980.
13. **Song, P. S. and Ou, C. N.,** Labeling of nucleic acids with psoralens, *Ann. N.Y. Acad. Sci.*, 346, 355, 1980.
14. **Kochevar, I., Gange, E., and William, R.,** Cutaneous photobiology, *Photochem. Photobiol.*, 37, 695, 1983.
15. **Rodighiero, G. and Dall'Acqua, F.,** Biochemical and medical aspects of psoralens, *Photochem. Photobiol.*, 24, 647, 1976.
16. **Musajo, L. and Rodighiero, G.,** Mode of photosensitizing action of furocoumarins, in *Photophysiology*, Vol. 7, Giese, A., Ed., Academic Press, New York, 1971, 115.
17. **Girotti, A. W.,** Mechanisms of photosensitization, *Photochem. Photobiol.*, 386, 745, 1983.
18. **Brendel, M. and Ruhland, A.,** Relationships between functionality and genetic toxicology of selected DNA-damaging agents, *Mutat. Res.*, 133, 51, 1984.
19. **Parsons, B. J.,** Psoralen photochemistry, *Photochem. Photobiol.*, 326, 813, 1980.
20. **Kornhauser, A.,** Molecular aspects of phototoxicity, *Ann. N.Y. Acad. Sci.*, 346, 398, 1980.
21. **Moysan, A., Gaboriau, F., Vigny, P., Voituriez, L., and Cadet, J.,** Chemical structure of 3-carbethoxypsoralen-DNA photoadducts, *Biochimie*, in press.
22. **Gaboriau, F., Vigny, P., Cadet, J., Voituriez, L., and Bisagni, E.,** Photoreaction of monofunctional 3-carbethoxypsoralen with DNA identification and conformational study of the predominant *cis-syn* furan side monoadduct to thymine, *Photochem. Photobiol.*, 45, 199, 1987.
23. **Fitzpatrick, T. B., Parrish, J. A., and Pathak, M. A.,** Photobiology of vitiligo (idiopathic leukodermal), in *Sunlight and Man: Normal and Abnormal Photobiologic Responses*, Fitzpatrick, T. B., Pathak, M. A., Haber, L. C., Seiji, M., and Kukita, A., Eds., University of Tokyo Press, Tokyo, 1974, 783.
24. **El Mofty, A. M.,** *Vitiligo and Psoralens*, Oxford, Pergamon Press, New York, 1968.
25. **Kenney, J. A., Jr.,** Vitiligo treated by psoralens. A long-term follow-up study of the permanency of repigmentation, *Arch. Dermatol.*, 103(5), 475, 1971.
26. **Punshi, S. K.,** Vitiligo, *Q. Med. Rev.*, 30(4), 1, 1979.
27. **Musajo, L. and Rodighiero, G.,** The skin-photosensitizing furocoumarins, *Experientia*, 28, 153, 1962.
28. **Ivie, G. W., Holt, D. L., and Marcellus, C.,** Natural toxicants in human foods, psoralens in raw and cooked parsnip root, *Science*, 213, 909, 1981.
29. **Pathak, M. A.,** Phytophotodermatitis, in *Sunlight and Man: Normal and Abnormal Photobiologic Responses*, Fitzpatrick, T. B., Pathak, M. A., Haber, L. C., Seiji, M., and Kukita, A., Eds., University of Tokyo Press, Tokyo, 1974, 495.
30. **Fowlks, W. L.,** The chemistry of psoralen, *J. Invest. Dermatol.*, 32, 249, 1959.
31. **Phyladelphy, A.,** Zur Atiologie der Wiesenpflanzen - (Bade) Dermatitis, *Dermatol. Wochschr.*, 92, 713, 1931.
32. **Walter, J. F., Kelsey, W. H., Voorhees, J. J., and Duell, E. A.,** Psoralen plus black light inhibits epidermal DNA synthesis, *Arch. Dermatol.*, 107(6), 861, 1973.
33. **Parrish, J. A., Fitzpatrick, T. B., Tanenbaum, L., and Pathak, M. A.,** Photochemotherapy of psoriasis with oral methoxsalen and long wave ultraviolet light, *N. Engl. J. Med.*, 291, 1207, 1974.
34. **Wolff, K. and Honigsmann, H.,** Clinical aspects of photochemotherapy, *Pharmacol. Ther.*, 12, 381, 1981.
35. **Melski, J. W., Tanenbaum, L., Parrish, J. A., Fitzpatrick, T. B., Bleich, H. L., and 28 participating investigators,** Oral methoxsalen photochemotherapy for the treatment of psoriasis: a cooperative clinical trial, *J. Invest. Dermatol.*, 68, 328, 1977.
36. **Nietsche, U. B.,** Photochemotherapy of psoriasis, *Int. J. Dermatol.*, 17(2), 149, 1978.
37. **Parrish, J. A., Le Vine, M. J. and Fitzpatrick, T. B.,** Oral methoxsalen photochemotherapy of psoriasis and mycosis fungoides, *Int. J. Dermatol.*, 19(7), 379, 1980.

38. **Bryant, B. G.**, Treatment of psoriasis, *Am. J. Hosp. Pharm.*, 37(6), 814, 1980.
39. **Wolff, K., Honigsman, H. Gschnait, F., and Konrad, K.**, Photochemotherapie bei Psoriasis, *Dtsch. Med. Wochenschr.*, 100, 2471, 1975.
40. **Hofmann, C., Pelwig, G., and Braun-Falco, O.**, Klinische Erfahrungen mit der 8-Methoxypsoralen-UVA-Therapie (Photochemotherapie) bei Psoriasis, *Hautarzt*, 27, 588, 1976.
41. **Pullmann, H., Zingsheim, M., Steigleder, G. K., and Orfanos, C. E.**, PUVA- und Anthralintherapie der Psoriasis, ein Klinischer, histologischer und autoradiographischer Vergleich, *Z. Hautkr.*, 51, 861, 1976.
42. **Thivolet, J., Ortonne, J. P., and Chouvet, B.**, La photochimiotherapie du psoriasis. Technique et resultats, *Lyon Med.*, 237, 87, 1977.
43. **Morison, W. L., Parrish, J. A. and Fitzpatrick, T. B.**, Controlled study of PUVA and adjunctive topical therapy in the management of psoriasis, *Br. J. Dermatol.*, 98, 125, 1978.
44. **Epstein, J. H.**, Photomedicine, in *The Science of Photobiology*, Smith, K. C., Ed., Plenum Press, New York, 1977, 175.
45. **Fischer, T., Alsins, J., Claesson, S. and Juhlin, L.**, in *Photochemotherapy: Basis, Technique and Side Effects*, Jung, E. G., Ed., Schattauer Verlag, Stuttgart, 1976, 91.
46. **Wolff, K., Honigsmann, H., Gschnait, F., and Konrad, K.**, in *Photochemotherapy: Basis, Technique and Side Effects*, Jung, E. G., Ed., Schattauer Verlag, Stuttgart, 1976, 81.
47. **Anderson, T. F. and Voorhees, J. J.**, Psoralen photochemotherapy of cutaneous disorders, *Annu. Rev. Pharmacol. Toxicol.*, 20, 235, 1980.
48. **Parrish, J. A., Le Vine, M. J., and Fitzpatrick, T. B.**, Oral methoxsalen photochemotherapy of psoriasis and mycosis fungoides, *Int. J. Dermatol.*, 19(7), 379, 1980.
49. **Honigsmann, A. K., Konrad, K., Gschnait, F., et al.**, (Abstr.) 7th Int. Cong. Photobiology, Rome, August 29 to September 3, 1976.
50. **Roenigk, H. H., Jr.**, (Abstr.) 7th Int. Cong. Photobiology, Rome, August 29 to September 3, 1976.
51. **Vonderheid, E. C.**, Evaluation and treatment of mycosis fungoides lymphoma, *Int. J. Dermatol.*, 19(4), 182, 1980.
52. **Hofmann, C., Burg, G., Plewig, G., Braun, G., and Falco, O.**, Photochemotherapie Kutaner Lymphome, *Dtsch. Med. Wochenschr.*, 102, 675, 1977.
53. **Gschnait, F., Honigsmann, H., Konrad, K., and Wolff, K.**, paper presented at the 75th Cong. Photobiology, Rome, August 29 to September 3, 1976.
54. **Uematsu, S. and Mizuno, N.**, paper presented at the 7th Int. Cong. Photobiology, Rome, August 29 to September 3, 1976.
55. **Scheen, S. R., III, Connolly, S. M., and Dicken, C. H.**, Actinic prurigo, *J. Am. Acad. Dermatol.*, 5(2), 183, 1981.
56. **Willis, I.**, Photochemotherapy: a new promising chemical derivative, *Arch. Dermatol.*, 272, 251, 1982.
57. **Pathak, M. A., Fellman, J. H., and Kaufman, K. D.**, The effects of structural alterations on the erythemal activity of furocoumarins: psoralens, *J. Invest., Dermatol.*, 35, 165, 1960.
58. **Nordlund, J. J., Ackles, A. E., and Traynor, F. F.**, The proliferative and toxic effects of ultraviolet light and inflammation on epidermal pigment cells, *J. Invest. Dermatol.*, 77(4), 361, 1981.
59. **Pathak, M. A., Kramer, D. M., and Fitzpatrick, T. B.**, Photobiology and photochemistry of furocoumarins (psoralens), in *Sunlight and Man: Normal and Abnormal Photobiologic Responses*, Fitzpatrick, T. B., Pathak, M. A., Haber, L. C., Seiji, M., and Kukita, A., Eds., University of Tokyo Press, Tokyo, 1974, 335.
60. **Granstein, R. D. and Sober, A. J.**, Drug- and heavy metal-induced hyperpigmentation, *J. Am. Acad. Dermatol.*, 5(1), 1, 1981.
61. **Jimbow, K., Kaidbey, K. H., Pathak, M. A., Parrish, J. A., Kligman, A. M., and Fitzpatrick, T. B.**, Melanin pigmentation stimulated by UV-B, UV-A, and psoralens, *J. Invest. Dermatol.*, 62, 548, 1974.
62. **Ortonne, J. P., Perrot, H., Schmitt, D., and Bioulac, P.**, Cutaneous hypermelanosis and intramelanotic lipid droplets, *Arch. Dermatol.*, 116(3), 301, 1980.
63. **Costa-Martins, J. E., Pozetti, G. L., and Sodre, M.**, Effects of psoralen and bergapten on irradiated skin, *Int. J. Dermatol.*, 13(3), 124, 1974.
64. **Kaidbey, K. H. and Kligman, A. M.**, Photopigmentation with trioxsalen, *Arch. Dermatol.*, 109(5), 674, 1974.
65. **Lerman, S.**, Psoralens and ocular effects in man and animals: in vivo monitoring of human ocular and cutaneous manifestations, *Natl. Cancer Inst. Monogr.*, 66, 227, 1984.
66. **Lerman, S.**, An experimental and clinical evaluation of lens transparency and aging, *J. Gerontol.*, 38(3), 293, 1983.
67. **Lerman, S.**, Ocular phototoxicity and PUVA therapy: an experimental and clinical evaluation, *J. Natl. Cancer Inst.*, 69, 287, 1982.
68. **Lerman, S.**, Potential ocular complications of psoralen-UV-A therapy, *Derm. Beruf. Umwelt*, 28(1), 5, 1980.

69. **Freeman, R. G. and Troll, D.,** Photosensitization of the eye by 8-methoxypsoralen, *J. Invest. Dermatol.,* 53(6), 449, 1969.

70. **Lerman, S. and Borkman, R. F.,** Photochemistry and lens aging, in *Gerontological Aspects of Eye Research,* Hockwin, O., Ed., S. Karger, Basel, 1978, 154.

71. **Parrish, J. A., Chylack, L. T., Jr., Woehler, M. E., Cheng, H. M., Pathak, M. A., Morison, W. L., Krugler, J., and Nelson, W. F.,** Dermatological and ocular examinations in rabbits chronically photosensitized with methoxsalen, *J. Invest. Dermatol.,* 73(3), 250, 1979.

72. **Glew, W. B. and Nigra, T. P.,** Psoralens and ocular effects in humans, *Natl. Cancer Inst. Monogr.* 66, 235, 1984.

73. **Stern, R. S., Parrish, J. A., and Fitzpatrick, T. B.,** Ocular findings in patients treated with PUVA, *J. Invest. Dermatol.,* 85, 269, 1985.

74. **Becker, S. W., Jr.,** Detection of photosensitizers in tissue, *Arch. Dermatol.,* 92(4), 457, 1965.

75. **Grube, D. D., Ley, R. D., and Fry, R. J. M.,** Photosentizing effects of 8-methoxypsoralen on the skin of hairless mice. II. Strain and spectral differences for tumorigenesis, *Photochem. Photobiol.,* 25, 269, 1977.

76. **Szafarz, D., Zajdela, F., Bornecque, C., and Barat, N.,** Evaluation of DNA crosslinks and monoadducts in mouse embryo fibroblasts after treatment with mono- and bifunctional furocoumarins and 365 nm, (UVA) irradiation. Possible relationship to carcinogenicity, *Photochem. Photobiol.,* 38, 557, 1983.

77. **Griffin, A. C., Hakim, R. E., and Knox, J. M.,** The wavelength effect upon erythemal and carcinogenic response in psoralen treated mice, *J. Invest. Dermatol.,* 31, 289, 1958.

78. **Hakim, R. E., Griffin, A. C., and Knox, J. M.,** Erythema and tumor formation in methoxsalen-treated mice exposed to fluorescent light, *Arch. Dermatol.,* 82, 572, 1960.

79. **Roelandts, R.,** Mutagenicity and carcinogenicity of methoxsalen plus UV-A, *Arch. Dermatol.,* 120(5), 662, 1984.

80. **Cartwright, L. E. and Walter, J. F.,** Psoralen-containing sunscreen is tumorigenic in hairless mice, *J. Am. Acad. Dermatol.,* 8, 830, 1983.

81. **Van der Leun, J. C.,** UV-carcinogenesis, *Photochem. Photobiol.,* 39(6), 861, 1984.

82. **Roelandts, R.,** Mutagenicity and carcinogenicity of methoxsalen plus UV-A, *Arch. Dermatol.,* 120(5), 662, 1984.

83. **Grekin, D. A. and Epstein, J. H.,** Psoralens, UVA (PUVA) and photocarcinogenesis, *Photochem. Photobiol.,* 33(6), 957, 1981.

84. **Bridges, B. A.,** Possible long-term hazards of photochemotherapy with psoralens and near ultraviolet light, *Clin. Exp. Dermatol.* 3(4), 349, 1978.

85. **Koch, W. H.,** Psoralen photomutagenic specificity in *Salmonella typhimurium, Mutat. Res.,* 160, 195, 1986.

86. **Verturini, S., Tamaro, M., Monti-Bragadin, C., and Carlassare, F.,** Mutagenicity in *Salmonella typhimurium* of some angelicin derivatives proposed as new monofunctional agents for the photochemotherapy of psoriasis, *Mutat. Res.,* 88, 17, 1981.

87. **Harter, M. L., Felkner, I. C., Mantulin, W. W., McInturff, D. L., Marx, J. N., and Song, P. S.,** Excited states and photobiological properties of potential DNA cross-linking agents, the benzodipyrones, *Photochem. Photobiol.,* 20, 407, 1974.

88. **Roelandts, R.,** Mutagenicity and carcinogenicity of methoxsalen plus UV-A, *Arch. Dermatol.,* 120(5), 662, 1984.

89. **De Mol, N. J., Beijersbergen van Henegouwen, G. M. J., Mohn, G. R., Glickman, B. W., and Van Kleef, P. M.,** On the involvement of singlet oxygen in mutation induction by 8-methoxypsoralen and UVA irradiation in *Escherichia coli* K-12, *Mutat. Res.,* 82, 23, 1981.

90. **Ormsby, O. S.,** *Diseases of the Skin,* Lea & Febiger, Philadelphia, 1915, 309.

91. **Halprin, K. M., Comerford, M., and Taylor, J. R.,** Skin cancer in patients treated with 8-methoxypsoralen plus longwave ultraviolet radiation, *Natl. Cancer Inst. Monogr.,* 66, 185, 1984.

92. **Turro, N. J.,** *Molecular Photochemistry,* Benjamin, New York, 1966.

93. **Calvert, J. G. and Pitts, J. N., Jr.,** *Photochemistry,* John Wiley & Sons, New York, 1966.

94. **Lamola, A. A.,** Fundamental aspects of the spectroscopy and photochemistry of organic compounds; electronic energy transfer in biologic systems; and photosensitization, in *Sunlight and Man: Normal and Abnormal Photobiologic Responses,* Fitzpatrick, T. B., Pathak, M. A., Haber, L. C., Seiji, M. and Kukita, A., Eds., University of Tokyo Press, Tokyo, 1974, 17.

95. **Musajo, L., Rodighiero, G., Colombo, G., Torlone, V. and Dall'Acqua, F.,** Photosensitizing furocoumarins: interaction with DNA and photo-inactivation of DNA containing viruses, *Experientia,* 21, 22, 1965.

96. **Musajo, L., Rodighiero, G, and Dall'Acqua, F.,** Evidences of a photoreaction of the photosensitizing furocoumarins with DNA and with pyrimidine nucleosides and nucleotides, *Experientia,* 21, 24, 1965.

97. **Musajo, L., Rodighiero, G., Breccia, A., Dall'Acqua, F., and Malesani, G.,** Skin-photosensitizing furocoumarins: photochemical interaction between DNA and -O^{14}CH^3 bergapten (5-methoxypsoralen), *Photochem. Photobiol.*, 5, 739, 1965.

98. **Musajo, L. and Rodighiero, G.,** The mechanism of action of the skin photosensitizing furocoumarins, *Acta Derm. Venereol. (Stockholm)*, 47(5), 298, 1967.

99. **Krauch, C. H., Karmer, D. M., and Wacker, A.,** Zum Wirkungsmechanismus Photodynamischer Furocoumarin Photoreaktion von Psoralen-(4-^{14}C) mit DNS, RNS, Homopolynucleotiden und Nucelosiden, *Photochem. Photobiol.*, 6, 341, 1967.

100. **Davies, R. J. H.,** The ultraviolet photochemistry of nucleic acids and their components, *Photochem. Photobiol.*, 31, 623, 1980.

101. **Ou, C. N. and Song, P. S.,** Photobinding of 8-methoxypsoralen to transfer RNA and 5-fluorouracil-enriched transfer RNA, *Biochemistry*, 17, 1054, 1978.

102. **Hochkeppel, H. K. and Gordon, J.,** Evidence for cross-linking of polyribonucleotides with 4'-amino-methyl-4,5',8-trimethylpsoralen hydrochloride, *Biochemistry*, 18, 2905, 1979.

103. **Palu, G., Palumbo, M., Cusinato, R., Meloni, G. A., and Marciani-Magno, S.,** Antiviral properties of psoralen derivatives: a biological and physico-chemical investigation, *Biochem. Pharmacol.*, 33(21), 3451, 1984.

104. **Nakashima, K., LaFiandra, A. J., and Shatkin, A. J.,** Differential dependence of reovirus-associated enzyme activities on genome RNA as determined by psoralen photosensitivity, *J. Biol. Chem.*, 254(16), 8007, 1979.

105. **Hradecna, Z. and Kittler, L.,** Photobiology of furocoumarins. Various types of crosslinking with DNA and their interference with the development of lambda phage, *Acta Virol. (Prague)*, 26(5), 305, 1982.

106. **Cole, R. S.,** Inactivation of *Escherichia coli* , F' episomes at transfer, and bacteriophage lambda by psoralen plus 360-nm light: significance of deoxyribonucleic acid cross-links, *J. Bacteriol.*, 107(3), 846, 1971.

107. **Moore, S. P., Blount, H., and Coohill, T. P.,** SV40 induction from a mammalian cell line by ultraviolet radiation and the photosensitizers 8-methoxypsoralen and angelicin, *Photochem. Photobiol.*, 37, 665, 1983.

108. **Colombo, G.,** Phostosensitization of sea-urchin sperm to long-wave ultraviolet light by psoralen, *Exp. Cell Res.*, 48(1), 167, 1967.

109. **Diener, T. O., McKinley, M. P., and Prusiner, S. B.,** Viroids and prions, *Proc. Natl. Acad. Sci. U.S.A.*, 79(17), 5220, 1982.

110. **Seki, T., Nozu, K., and Kondo, S.,** Differential causes of mutations and killing in *Escherichia coli* after psoralen plus light treatment: monoadducts and cross-links, *Photochem. Photobiol.*, 27, 19, 1978.

111. **Ames, B., Durston, W. E., Yamasaki, E., and Lee, F. D.,** Carcinogens are mutagens: A simple test system combining liver homogenates for activation and bacteria for detection, *Proc. Natl. Acad. Sci. U.S.A.*, 70, 2281, 1973.

112. **Waring, M. J.,** DNA modification and cancer, *Annu. Rev. Biochem.*, 50, 159, 1981.

113. **Averbeck, D. and Moustacchi, E.,** 8-Methoxypsoralen plus 365 nm light effects and repair in yeast, *Biochim. Biophys. Acta*, 395(4), 393, 1975.

114. **Baden, H. P., Parrington, J. M., Delhanty, J. D., and Pathak, M. A.,** DNA synthesis in normal and xeroderma pigmentosum fibroblasts following treatment with 8-methoxypsoralen and long wave ultraviolet light, *Biochim. Biophys. Acta*, 262(3), 247, 1972.

115. **Ou, C. N., Tsai, C. H., Tapley, K. J., Jr., and Song, P. S.,** Photobinding of 8-methoxypsoralen and 5,7-dimethoxycoumarin to DNA and its effect on template activity, *Biochemistry*, 17, 1047, 1978.

116. **Nielsen, P. E. and Linnane, W. P., 3rd,** Differentiated inhibition of DNA, RNA and protein synthesis in L1210 cells by 8-methoxypsoralen, *Biochem. Biophys. Res. Commun.*, 112(3), 965, 1983.

117. **Hyodo, M., Fujita, H., Suzuki, K., Yoshino, K., Matsuo, I., and Ohkido, M.,** DNA replication and cell-cycle progression of cultured mouse FM3a cells after treatment with 8-methoxypsoralen plus near UV-radiation, *Mutat. Res.*, 94(1), 199, 1982.

118. **Heimer, Y. M., Ben-Hur, E., and Riklis, E.,** Photosensitized inhibition of nitrate reductase induction by 4,5',8-trimethylpsoralen and near ultraviolet light, *Biochim. Biophys. Acta*, 519(2), 499, 1978.

119. **Heimer, Y. M., Ben-Hur, E., and Riklis, E.,** Psoralen and near ultraviolet light: a probe for study of control of protein synthesis (letter), *Nature (London)*, 268(5616), 170, 1977.

120. **Bordin, F., Baccichetti, F., Bevilacqua, R., and Musajo, L.,** Inhibition of protein synthesis in Ehrlich ascites tumor cells by irradiation (365 nm) in the presence of skin-photosensitizing furocoumarins, *Experientia*, 29(3), 272, 1973.

121. **Musajo, L. and Baccichetti, F.,** Protection against Graffi's leukemia in mice treated with leukemic cells photo-inactivated by psoralen, *Eur. J. Cancer*, 8(4), 397, 1972.

122. **Woo, T. Y., Wong, R. C., Wong, J. M., Anderson, T. F., and Lerman, S.,** Lenticular psoralen photoproducts and cataracts of a PUVA-treated psoriatic patient, *Arch. Dermatol.*, 121, 1307, 1985.

123. **Lerman, S., Megaw, J., Gardner, K., Takei, Y., Franks, Y., and Gammon, A.,** Photobinding of ^{3}H-8-methoxypsoralen to monkey intraocular tissues, *Invest. Ophthalmol. Vis. Sci.*, 25(11), 1267, 1984.

124. **Grossweiner, L. I.,** Photochemistry of proteins: a review, *Curr. Eye Res.*, 3(1), 137, 1984.

125. **Fredericksen, S. and Hearst, J. E.,** Binding of 4'-aminomethyl-4,5',8-trimethylpsoralen to DNA, RNA and protein in HeLa cells and *Drosophila* cells, *Biochim. Biophys. Acta*, 563, 343, 1979.

126. **Meffert, H., Diezel, W., Gunther, W., and Sonnichsen, N.,** Photochemotherapy of psoriasis using 8-methoxypsoralen and UVA. II. Binding of the photosensitizator to protein, *Dermatol. Monatsschr.*, 162, 887, 1976.

127. **Pathak, M. A. and Kramer, D. M.,** Photosensitization of skin in vivo by furocoumarins (psoralens), *Biochim. Biophys. Acta*, 195(1), 197, 1969.

128. **Engel, P. F. and Wulf, H. C.,** Localization of radioactivity in rat organs after oral administration of tritiated 8-methoxypsoralen in therapeutic doses, *Arch. Dermatol. Res.*, 273, 71, 1982.

129. **Zarebska, Z., Jarzabek-Chorzelska, M., Rzesa, G., Glinski, W., Pawinska, M., Chorzelski, T., and Jablonska, S.,** Detection of DNA-psoralen photoadducts *in situ*, *Photochem. Photobiol.*, 39, 307, 1984.

130. **Laskin, J. D., Lee, E., Yurkow, E. J., Laskin, D. L., and Gallo, M. A.,** A possible mechanism of psoralen phototoxicity not involving direct interaction with DNA, *Proc. Natl. Acad. Sci. U.S.A.*, 82, 6158, 1985.

131. **Bredberg, A. Lambert, B., Swanbeck, G., and Thyresson-Hok, M.,** The binding of 8-methoxypsoralen to nuclear DNA of UVA irradiated human fibroblasts *in vitro*, *Acta Derm. Venereol. (Stockholm)*, 57, 389, 1977.

132. **Bertaux, B., Dubertret, L., and Moreno, G.,** Autoradiographic localization of 8-methoxypsoralen in psoriasis skin in vitro, *Acta Derm. Venereol. (Stockholm)*, 61(6), 481, 1981.

133. **Meffert, H. von, Barthelmes, H., Metz, D., and Sonnichsen, N.,** Fotochemotherapie mit 8-Methoxypsoralen und UVA. III. Fluoreszenzmikroskopische Untersuchung der Bindung und des Transportes von 8-MOP in der Menschlichen Haut, (Photochemotherapy using 8-methoxypsoralen and UV light. III. Fluorescence microscopic study of the binding and transport of 8-MOP in human skin), *Dermatol. Monatsschr.*, 163, 619, 1977.

134. **Cech, T., Pathak, M. A., and Biswas, R. K.,** An electron microscopic study of the photochemical cross-linking on DNA in guinea-pig epidermis by psoralen derivatives, *Biochim. Biophys. Acta*, 562, 342, 1979.

135. **Ali, R., and Agarwala, S. C.,** In vitro and in vivo effect of normal and irradiated psoralen on glucose oxidation of brain and liver, *Enzyme*, 18(6), 321, 1974.

136. **Toda, K., Shono, S., and Imura, M.,** A trial of a method to trace metabolic pathways of nucleic acids in keratinocytes, *Curr. Probl. Dermatol.*, 10, 219, 1980.

137. **Song, P. S.,** Photoreactive states of furocoumarins, *Natl. Cancer Inst. Monogr.*, 66, 15, 1984.

138. **Song, P. S., Harter, M. L., Moore, T. A., and Herndon, W. C.,** Luminescence spectra and photocycloaddition of the excited courmarins to DNA bases, *Photochem. Photobiol.*,m 14, 521, 1971.

139. **Ray, N. K. and Ahuja, V. K.,** Molecular orbital study of the photoreactivity of triplet coumarins, *Photochem. Photobiol.*, 17, 347, 1973.

140. **Bevilacqua, R. and Bordin, F.,** Photo-C_4-cycloaddition of psoralen and pyrimidine bases: effect of oxygen and paramagnetic ions, *Photochem. Photobiol.*, 17, 191, 1973.

141. **Kanne, D., Straub, K., Hearst, J. E., and Rapoport, H.,** Isolation and characterization of pyrimidine-psoralen-pyrimidine photodiadducts from DNA., *J. Am. Chem. Soc.*, 104, 6754, 1982.

142. **Kanne, D., Straub, K., Rapoport, H., and Hearst, J. E.,** Psoralen-deoxyribonucleic acid photoreaction. Characterization of the monoaddition products from 8-methoxypsoralen and 4,5',8-trimethylpsoralen, *Biochemistry*, 21, 861, 1982.

143. **Gasparro, F. P., Bagel, J., and Edelson, R. L.,** HPLC analysis of 4',5'-monoadduct formation in calf thymus DNA and synthetic polynucleotides treated with UVA and 8-methoxypsoralen, *Photochem. Photobiol.*, 42, 95, 1985.

144. **Beaumont, P. C., Rodgers, M. A. J., Parsons, B. J., and Philips, G. O.,** Singlet oxygen production by some furocoumarin derivatives in the presence of DNA: time resolved luminescence measurements, *Photochem. Photobiol.*, 42, 605, 1985.

145. **Kan, R. O.,** *Organic Photochemistry*, McGraw-Hill, New York, 1966, 154.

146. **Zhang, M. and An, J.,** Photocycloaddition between 8-methoxypsoralen and olefins, *Huaxue Xuebao*, 41, 182, 1983.

147. **Kittler, L. K., Midden, W. R., and Wang, S. Y.,** Interactions of furocoumarins with subunits of cell constituents. Photoreaction of fatty acids and aromatic amino acids with TMP and 8-MOP, *Stud. Biophys*, 114, 139, 1986. Furocoumarins form covalent adducts with fatty acids. Implications for the design of noncarcinogenic psoralens, *Photochem. Photobiol.*, 43S, 17S, 1986.

148. **Kittler, L., Specht, K. G., and Midden, W. R.,** A new biological target of furocoumarins: photochemical formation of covalent adducts with unsaturated fatty acids, *Photochem. Photobiol.*, in press, 1988.

149. **Kittler, L., Midden, W. R., and Wang, S. Y.,** Photoreactions of lipids sensitized by furocoumarins, *Photochem. Photobiol.*, 37S, S16, 1983.

150. **Wang, S. Yi. and Midden, W. R.,** Photodynamic action on nucleic acids and their components, *Stud. Biophys.*, 94, 7, 1983.

151. **Cadet, J., Decarroz, C., Wang, S. Y., and Midden, W. R.,** Mechanisms and products of photosensitized degradation of nucleic acids and related model compounds, *Isr. J. Chem.*, 23, 420, 1983.

152. **Midden, W. R.,** Traces of oxygen were removed by scrubbing the gas with an oxygen scavenging preparation of vanadium over zinc amalgam, personal observation.

153. **Krauch, C. H., Farid, S., Kraft, S., and Wacker, A.,** Zum Wirkungsmechanismus photodynamischer Furocoumarine, *Biophysik*, 2, 301, 1965.

154. **Oginsky, E. L., Green, G. S., Griffith, G., and Fowlks, W. L.,** Lethal photosensitization of bacteria with 8-methoxypsoralen to long wavelength ultraviolet radiation, *J. Bacteriol.*, 78, 821, 1959.

155. **Musajo, L., Rodighiero, G., Caporale, G., Dall'Acqua, F., Marciani, S., Bordin, F., Bacchichietti, F., and Bevilacqua, R.,** Photoreactions between skin-photosensitizing furocoumarins and nucleic acids, in *Sunlight and Man: Normal and Abnormal Photobiological Responses*, Fitzpatrick, T. B., Pathak, M. A., Haber, L. C., Seiji, M., and Kukita, A., Eds., University of Tokyo Press, Tokyo, 1974, 369.

156. **Joshi, P. C., Wang, S. Y., Midden, W. R., Voiturez, L., and Cadet, J.,** Heterodimers of 8-methoxypsoralen and thymine, *Photobiochem. Photobiophys.*, 8, 51, 1984.

157. **Cadet, J., Voitiuriez, L., Ulrich, J., Joshi, P. C., and Wang, S. Y.,** Isolation and characterization of the monoheterodimers of 8-methoxypsoralen and thymidine involving the pyrone moiety, *Photobiochem. Photobiophys.*, 8, 35, 1984.

158. **Downes, C. P. and Michell, R. H.,** Inositol phospholipid breakdown as a receptor-controlled generator of second messengers, in *Molecular Aspects of Cellular Regulation*, Vol. 4, *Molecular Mechanisms of Transmembrane Signaling*, Cohen, P. and Houslay, M. D., Eds., Elsevier, New York, 1985, 3.

159. **Nishizuka, Y.,** Studies and perspectives of protein kinase C, *Science*, 233, 305, 1986.

160. **Ruzicka, T., Walter, J. F., and Printz, M. P.,** Altered arachidonic-acid metabolism in UV irradiated hairless mouse skin, 43rd Annual Meeting of The Society for Investigative Dermatology, Inc., Washington, D.C., U.S.A., *J. Invest. Dermatol.*, 78(4), 357, 1982.

161. **Cantieri, J. S., Graff, G., and Goldberg, N. D.,** Cyclic GMP metabolism in psoriasis: activation of soluble epidermal guanylate cyclase by arachidonic acid and 12-hydroxy-5,8,10,14-eicosatetraenoic acid, *J. Invest. Dermatol.*, 74, 234, 1980.

162. **Albrightson, C. R., Fertel, R. H., Brown, B. V., Stephens, R.,** Psoralens increase the concentration of cyclic AMP in human cells *in vitro*, *J. Invest. Dermatol.*, 85, 264, 1985.

163. **Le Vine, M. J., McGilvray, N., and Baden, H. P.,** Effect of therapy on keratin polypeptide profiles of psoriatic epidermis, *Arch. Dermatol.*, 116(9), 1028, 1980.

164. **Lowe, N. J., Connor, M. J., Cheong, E. S., Akopiantz, P., and Breeding, J. H.,** Psoralen and ultraviolet A effects on epidermal ornithine decarboxylase induction and DNA synthesis in the hairless mouse, *Natl. Cancer Inst. Monogr.*, 66, 73, 1984.

165. **Connor, M. J. and Lowe, N. J.,** The induction of erythema, edema, and the polyamine synthesis enzymes ornithine decarboxylase and S-adenosyl-L-methionine decarboxylase in hairless mouse skin by psoralens and longwave ultraviolet light, *Photochem. Photobiol.*, 39(6), 787, 1984.

166. **Folsom, J., Gange, R. W., and Mendelson, I. R.,** Ornithine decarboxylase induction in psoralen-treated mouse epidermis used as a test of UV-A sunscreen potency, *Br. J. Dermatol.*, 108(1), 17, 1983.

167. **Gange, R. W., and DeQuoy, P.,** Ornithine decarboxylase and DNA synthesis in mouse skin treated with PUVA or anthracene + UVA, *Clin. Res.*, 28, 134A.

168. **Gange, R. W.,** Epidermal ornithine decarboxylase activity and thymidine incorporation following treatment with ultraviolet A combined with topical 8-methoxypsoralen or anthracene in the hairless mouse, *Br. J. Dermatol.*, 105, 247, 1981.

169. **Walter, J. F., Gange, R. W. and Mendelson, I. R.,** Psoralen-containing sunscreen induces phototoxicity and epidermal ornithine decarboxylase activity, *J. Am. Acad. Dermatol.*, 6(6), 1022, 1982.

170. **Ben-Hur, E. and Riklis, E.,** Inhibition of induced ornithine decarboxylase activity in Chinese hamster cells by gamma radiation, far ultraviolet light and psoralen plus near-ultraviolet light: a comparative study, *Int. J. Radiat. Biol.*, 39(5), 527, 1981.

171. **Ben-Hur, E., Heimer, Y. M., and Riklis, E.,** Gamma radiation inhibits the appearance of induced ornithine decarboxylase activity in Chinese hamster cells, *Int. J. Radiat. Biol.*, 39(5), 515, 1981.

172. **Heimer, Y. M. and Riklis, E.,** Inhibition by trioxsalen (psoralen) plus near-ultraviolet light of the induction of ornithine decarboxylase in Chinese-hamster cells, *Biochem. J.*, 183(1), 179, 1979.

173. **Van De Kerkhof, P. C. M., Fleuren, E., Van Rennes, H., and Mier, P. D.,** Metabolic changes in the psoriatic lesion during therapy, *Br. J. Dermatol.*, 110(4), 411, 1984.

174. **Rytter, M., Gast, W., Hofmann, C., and Barth, J.,** Binding of 8-methoxypsoralen to the membrane of human lymphocytes without UV-irradiation, *Dermatol. Monatsschr.*, 168(12), 814, 1982.

175. **Lange, B., Meffert, H., and Bohm, F.,** Low-level chemiluminescence measurement of the binding of 8-methoxypsoralen to protein and to lymphocytic surfaces (transl.), *Dermatol. Monatsschr.*, 166(9), 599, 1980.

176. **Wennersten, G.,** Membrane damage caused by 8-MOP and UVA-treatment of cultivated cells, *Acta Derm. Venereol. (Stockholm)*, 59(1), 21, 1979.

177. **Merkel, P. B., Nilsson, R., and Kearns, D. R.,** Deuterium effects of singlet oxygen lifetimes in solutions. A new test of singlet oxygen reactions, *J. Am. Chem. Soc.,* 94, 1030, 1972.

178. **Merkel, P. B. and Kearns, D. R.,** Radiationless decay of singlet molecular oxygen in solution. An experimental and theoretical study of electronic-to-vibrational energy transfer, *J. Am. Chem. Soc.,* 94, 7244, 1972.

179. **Scherwitz, C., Rassner, G., and Martin, R.,** Effects of 8-methoxypsoralen plus 365 nm UVA light on Candida albicans cells. An electron microscopic study, *Arch. Dermatol. Res.,* 263(1), 47, 1978.

180. **Danno, K., Takigawa, M., and Horio, T.,** Alterations in lectin binding to the epidermis following treatment with 8-methoxypsoralen plus longwave UV radiation, *J. Invest. Dermatol.,* 82(2), 176, 1984.

181. **Danno, K., Takigawa, M., and Horio, T.,** The alterations of keratinocyte surface and basement membrane markers by treatment with 8-methoxypsoralen plus longwave UV light, *J. Invest. Dermatol.,* 80(3), 172, 1983.

182. **Varga, J. M., Wiesehahn, G., Bartholomew, J. C., and Hearst, J. E.,** Dose-related effects of psoralen and ultraviolet light on the cell cycle of murine melanoma cells, *Cancer Res.,* 42, 2223, 1982.

183. **Coon, W. M., Wheatley, V. R., Herrmann, F., and Mandol, L.,** Free fatty acids of the skin surface and barrier zone in normal and abnormal keratinization, *J. Invest. Dermatol.,* 41, 259, 1963.

184. **Herrmann, F. J., Scher, R., Coon, W. M., and Mandol, L.,** The acid number of the lipids on the intact and the stripped skin surface in psoriatics, *J. Invest. Dermatol.,* 35, 47, 1960.

185. **Coon, W. M., Herrmann, F. J., and Mandol, L.,** Acid number of skin surface lipids in psoriasis, *Arch. Dermatol.,* 83, 119, 1961.

186. **Scher, R., Herrmann, F. J., Coon, W. M., and Mandol, L.,** The acid number of the lipids on the stripped skin surface of patients with psoriasis and other parakeratotic plaques, *Acta Derm. Venereol. (Stockholm),* 42, 363, 1962.

187. **Coon, W. M., Wheatley, V. R., Herrmann, F., and Mandol. L.,** Free fatty acids of the skin surface and barrier zone in normal and abnormal keratinization, *J. Invest. Dermatol.,* 41, 259, 1963.

188. **Summerly, R., Ilderton, E., and Gray, G. M.,** Possible defects in triacylglycerol and phosphatidylchloline metabolism in psoriatic epidermis, *Br. J. Dermatol.,* 99, 279, 1978.

189. **Muller, R. R. and Grossweiner, L. I.,** Dark membrane lysis and photosensitization by 3-carbethoxy-psoralen, *Photchem. Photobiol.,* 33(3), 399, 1981.

190. **Cerutti, P. A.,** Prooxidant states and tumor promotion, *Science,* 227, 375, 1985.

191. **Fantone, J. C. and Ward, P. A.,** Oxygen-derived radicals and their metabolites: relationship to tissue injury, in *Curr. Concepts,* UpJohn Co., Kalamazoo, Mich., 1985.

192. **Frankel, E. N.,** Chemistry of free radical and singlet oxidation of lipids, *Prog. Lipid Res.,* 23, 197, 1985.

193. **Pryor, W. A., Ed.,** *Free Radicals in Biology,* Vols. 1—6, Academic Press, New York, 1976—1986.

194. **Joshi, P. C. and Pathak, M. A.,** The role of active oxygen ($^1O_2$ and O_2) induced by crude coal tar and its ingredients used in photochemotherapy of skin diseases, *J. Invest. Dermatol.,* 82, 67, 1984.

195. **Pathak, M. A. and Joshi, P. C.,** The nature and molecular basis of cutaneous photosensitivity reactions to psoralens and coal tar, *J. Invest. Dermatol.,* Suppl. 66s, 1983.

196. **Pathak, M. A. and Joshi, P. C.,** Production of active oxygen species ($^1O_2$ and O_2) by psoralens and ultraviolet radiation (320—400 nm), *Biochim. Biophys. Acta,* 798(1), 115, 1984.

197. **Joshi, P. C. and Pathak, M. A.,** Production of singlet oxygen and superoxide radicals by psoralens and their biological significance, *Biochem. Biophys. Res. Commun.,* 112(2), 638, 1983.

198. **Salet, C., Moreno, G., and Vinzens, F.,** Photodynamic effects induced by furocoumarins on a membrane system. Comparison with hematoporphyrin, *Photochem. Photobiol.,* 36(3), 291, 1982.

199. **Matsuo, I., Yoshino, K., and Ohkido, M.,** Mechanism of skin surface lipid peroxidation, *Curr. Probl. Dermatol.,* 11, 135, 1983.

200. **Davidson, R. S. and Trethewey, K. R.,** Factors affecting dye-sensitized photo-oxygenation reactions, *J. Chem. Soc. Perkin Trans. 2,* 169, 1977.

201. **Meffert, H., Bohm, F., Roder, B., and Sonnichsen, N.,** Is stimulated singlet oxygen involved in the PUVA therapy effect? The effect of heavy water on membrane damage and formation of free radicals after the effect of 8-MOP and UVA on human lymphocytes, *Dermatol. Monatsschr.,* 168(6), 387, 1982.

202. **Glass, K. F., Pliquett, E., Grimm, A. Y., Potapenko, G. A., Abijev, D. I., Roscupkin, Ju, A., and Vladimirov, W.,** *Z. Karl Marx Univ. Leipzig Math. Naturwiss. R,* 28, 148, 1979.

203. **Sukhorukov, V. L., Potanenko, A. I., and Zakharova, V. A.,** Dark oxidation of unsaturated lipids and dihydroxyphenylalanine by photooxidized furocoumarins, *Vopr. Med. Khim.,* 29(5), 75, 1983.

204. **Potapenko, A. Y., Moshnin, M. V., Krasnovsky, A. A., Jr., and Sukhorukov, V. L.,** Dark oxidation of unsaturated lipids by the photo-oxidized 8-methoxypsoralen, *Z. Naturforsch. Teil C,* 37(1—2), 70, 1982.

205. **Wasserman, H. H. and Berdahl, D. R.,** The photooxidation of 8-methoxypsoralen, *Photochem. Photobiol.,* 35, 565, 1982.

206. **Carter, D. M., Jegasothy, B. V., and Condit, E. S.,** Defense of cutaneous cells against UV irradiation. II. Restricted photomediated binding of trimethylpsoralen to pigmented skin *in vivo, J. Invest. Dermatol.,* 60, 274, 1973.

207. **Lerman, S. and Borkman, R. F.,** A method for detecting 8-methoxypsoralen in the ocular lens, *Science,* 197, 1287, 1977.

208. **Lerman, S., Jocoy, M., and Borkman, R. F.,** Photosensitization of the lens by 8-methoxypsoralen, *Invest. Ophthalmol.,* 16, 1065, 1977.

209. **Lerman, S., Megaw, J., and Willis, I.,** Potential ocular complications from PUVA therapy and their prevention, *J. Invest. Dermatol.,* 74, 197, 1980.

210. **Zigler, J. S., Jr., and Goosey, J. D.,** Photosensitized oxidation in the ocular lens: evidence for photosensitizers endogenous to the human lens, *Photochem. Photobiol.,* 33, 869, 1981.

211. **Megaw, J., Lee, J., and Lerman, S.,** NMR analyses of tryptophan-8-methoxypsoralen photoreaction products formed in the presence of oxygen, *Photochem. Photobiol.,* 32, 265, 1980.

212. **Lerman, S., Megaw, J., and Gardner, K.,** Psoralen-long-wave ultraviolet therapy and human cataractogenesis, *Invest. Ophthalmol. Vis. Sci.,* 23(6), 801, 1982.

213. **Lerman, S., Megaw, J., and Willis, I.,** The photoreactions of 8-methoxypsoralen with tryptophan and lens proteins, *Photochem. Photobiol.,* 31(3), 235, 1980.

214. **Veronese, F. M., Schiavon, O., Bevilacqua, R., Bordin, F., and Rodighiero, G.,** The effect of psoralens and angelicins on proteins in the presence of UV-A radiation, *Photochem. Photobiol.,* 34, 351, 1981.

215. **Melo, T. de S., Morliere, P., Goldstein, S., Santus, R., Dubertret, L., and Lagrange, D.,** Binding of 5-methoxypsoralen to human serum low density lipoproteins, *Biochem. Biophys. Res. Commun.,* 120(2), 670, 1984.

216. **Artuc, M., Stuettgen, G., Schalla, W., Schaefer, H., and Gazith, J.,** Reversible binding of 5- and 8-methoxypsoralen to human serum proteins (albumin) and to epidermis in vitro, *Br. J. Dermatol.,* 101(6), 669, 1979.

217. **Veronese, F. M., Bevilacqua, R., Schiavon, O., and Rodighiero, G.,** Drug-protein interaction: plasma protein binding of furocoumarins, *Farmaco Ed. Sci.,* 34(8), 716, 1979.

218. **Veronese, F. M., Bevilacqua, R., Schiavon, O., and Rodighiero, G.,** The binding of 8-methoxypsoralen by human serum albumin, *Farmaco Ed. Sci.,* 33(9), 667, 1978.

219. **Schiavon, O., Benassi, C. A., and Veronese, F. M.,** The study of the furocoumarin-serum albumin complex by difference spectrophotometry and gel exclusion, *Farmaco Ed. Prat.,* 39(4), 109, 1984.

220. **Beaumont, P. C., Land, E. J., Navaratnam, S., Parsons, B. J., and Phillips, G. O.,** A pulse radiolysis study of the complexing of furocoumarins with DNA and proteins, *Biochim. Biophys. Acta,* 608(1), 182, 1980.

221. **Bickers, D. R. and Pathak, M. A.,** Psoralen pharmacology: studies on metabolism and enzyme induction, *Natl. Cancer Inst. Monogr.,* 66, 77, 1984.

222. **Fouin-Fortunet, H., Tinel, M., Descatoire, V., Letteron, P., Larrey, D., Geneve, J., and Pessayre, D.,** Inactivation of cytochrome P-450 by the drug methoxsalen, *J. Pharmacol. Exp. Ther.,* 236(1), 237, 1986.

223. **Bickers, D. R., Mukhtar, H., Molica, S. J., Jr., and Pathak, M. A.,** The effect of psoralens on hepatic and cutaneous drug metabolizing enzymes and cytochrome P-450, *J. Inst. Dermatol.,* 79(3), 201, 1982.

224. **Mandula, B. B., Pathak, M. A., Nakayama, Y., and Davidson, S. J.,** Induction of mixed-function oxidases in mouse liver by psoralens, *Br. J. Dermatol.,* 99(6), 687, 1978.

225. **Tsambaos, D., Vizethum, W., and Goerz, G.,** Effect of oral 8-methoxypsoralen on rat liver microsomal cytochrome P-450, *Arch. Dermatol. Res.,* 263(3), 339, 1978.

226. **Bevilacqua, R., Baccichetti, F., and Musajo, L.,** Effect of irradiation at 365 nm in the presence of psoralen on the oxygen consumption in rat liver mitochondria and in Ehrlich ascites tumor cells, *Ital. J. Biochem.,* 22(4), 160, 1973.

227. **Costa, C., De Antoni, A., Baccichetti, F., Vanzan, S., and Allegri, G.,** Tryptophan metabolism in animals with dermatitis. II. In guinea pigs, *Boll. Soc. Ital. Biol. Sper.,* 55(17), 1714, 1979.

228. **De Antoni, A., Baccichetti, F., Costa, C., Vanzan, S., and Allegri, G.,** Tryptophan metabolism in animals with dermatitis. I. In rats, *Boll. Sco. Ital. Biol. Sper.,* 55(17), 1707, 1979.

229. **De Antoni, A., Costa, C., Allegri, G., Baccichetti, F., and Vanzan, S.,** Effect of psoralen-induced photodermatitis on tryptophan metabolism in rats, *Chem. Biol. Interact.,* 34(1), 11, 1981.

230. **Vaatainen, N., Oikarinen, A., and Kuutti-Savolainen, E. R.,** The effects of long-term local PUVA treatment on collagen metabolism in human skin, *Arch. Dermatol. Res.,* 269(1), 99, 1980.

231. **Nilsson, R., Merkel, P. B., and Kearns, D. R.,** Unambiguous evidence for the participation of singlet oxygen ($^1\Delta$) in photodynamic oxidation of amino acids, *Photochem. Photobiol.,* 16, 117, 1972.

232. **Matheson, I. B. C. and Lee, J.,** Chemical reaction rates of amino acids with singlet oxygen, *Photochem. Photobiol.,* 29, 879, 1979.

233. **Veronese, F. M., Schiavon, O., Bevilacqua, R., and Rodighiero, G.,** Drug-protein interaction: 8-methoxy psoralen as photosensitizer of enzymes and amino acids, *Z. Naturforsch. Teil. C,* 34C(5—6), 392, 1979.

234. **Yoshikawa, K., Mori, N., Sakakibara, S., Mizuno, N., and Song, P. S.,** Photo-conjugation of 8-methoxypsoralen with proteins, *Photochem. Photobiol.,* 29, 1127, 1979.

235. **Yoshikawa, K., Sakakibara, S., Mori, N., Mizuno, N., and Song, P. S.,** Mechanism of photochemical reaction between 8-methoxypsoralen and protein (transl.), *Nippon Hifuka Gakkai Zasshi,* 89(6), 417, 1979.

236. **Veronese, F. M., Schiavon, O., Bevilacqua, R., Bordin, F., and Rodighiero, G.,** Photoinactivation of enzymes by linear and angular furocoumarins, *Photochem. Photobiol.,* 36(1), 25, 1982.

237. **Schiavon, O. and Veronese, F. M.,** Extensive crosslinking between subunits of oligomeric proteins induced by furocoumarins plus UV-A irradiation, *Photochem. Photobiol.,* 43, 243, 1986.

238. **Vedaldi, D., Dall'Acqua, F., Gennaro, A., and Rodighiero, G.,** Photosensitized effects of furocoumarins: the possible role of singlet oxygen, *Z. Naturforsch. Teil C,* 38C, 866, 1983.

239. **Granger, M., Toulme, F., and Helene, C.,** Photodynamic inhibition of *Escherichia coli* DNA polymerase I by 8-methoxypsoralen plus near ultraviolet irradiation, *Photochem. Photobiol.,* 36, 175, 1982.

240. **Ronfard-Haret and Bensasson,** unpublished.

241. **Granger, M. and Helene, C.,** Photoaddition of 8-methoxypsoralen to *E. coli* DNA polymerase I. Role of psoralen photoadducts in the photosensitized alterations of pol I enzymatic activities, *Photochem. Photobiol.,* 38, 563, 1983.

250. **Poppe, W. and Grossweiner, L. I.,** Photodynamic sensitization by 8-methoxypsoralen via the singlet oxygen mechanism, *Photochem. Photobiol.,* 22, 217, 1975.

251. **Knox, C. N., Land, E. J., and Truscott, T. G.,** Singlet oxygen generation by furocoumarin triplet states. I. Linear furocoumarins (psoralens), *Photochem. Photobiol.,* 43, 359, 1986.

252. **Singh, H. and Vadsaz, J. A.,** A major reactive species in the furocoumarin photosensitized inactivation of *E. coli* ribosomes, *Photochem. Photobiol.,* 28, 539, 1978.

253. **Singh, H., Bishop, J., and Merritt, J.,** Singlet oxygen and ribosomes: inactivation and sites of damage, *J. Photochem.,* 25, 295, 1984.

254. **Kraljic, I.,** *Proc. 3rd Tihany Symp. Radiation Chemistry,* Dobo, J. and Hedvig, P., Eds. Akademiai Kiado, Budapest, 1971, 1405.

255. **Dorfman, L. M. and Adams, G. E.,** Reactivity of the Hydroxyl Radical in Aqueous Solutions, NSRDS-NBS 46, U.S. Government Printing Office, Washington, D.C., 1973.

256. **Rizzuto, F. and Spikes, J. D.,** The eosin-sensitized photooxidation of substituted phenylalanines and tyrosines, *Photochem. Photobiol.,* 25, 465, 1977.

257. **McCord, J. M.,** Free radicals and inflammation: protection of synovial fluid by superoxide dismutase, *Science,* 185, 529, 1974.

258. **Schiavon, O., Simonic, R., Ronchi, S., Bevilacqua, R., and Veronese, F. M.,** Modification of RNAse A by near UV irradiation in the presence of psoralen, *Photochem. Photobiol.,* 39(1), 25, 1984.

259. **Midden, W. R.,** unpublished observation.

260. **Craw, M., Chedekel, M. R., Trustcott, T. G., and Land, E. J.,** The photochemical interaction between the triplet state of 8-methoxypsoralen and the melanin precursor L-DOPA, *Photochem. Photobiol.,* 39(2), 155, 1984.

261. **Prota, G.,** Recent advances in the chemistry of melanogenesis in mammals, *J. Invest. Dermatol.,* 75, 122, 1980.

262. **Musajo, L., Rodighiero, G., Caporale, G., et al.,** La fotoreazione tra bergaptene e flavin-mononucleotide, *Gazz. Chim. Ital.,* 94, 1054, 1964.

263. **Rodighiero, G., Musajo, L., Fornasiero, U., Caporle, G., Malesani, G., and Chiarelotto, G.,** La fotoreazione tra xantotossina e flavin-mononucleotide, *Gazz. Chim. Ital.,* 94, 1084, 1964.

264. **Azzone, G. F., Rodighiero, G., Musajo, L., Dall'Acqua, F., Fornasiero, U., and Malesani, G.,** Richerche enzimologiche sui flavin foto-composti, *Gazz. Chim. Ital.,* 94, 1101, 1964.

265. **Rodighiero, G., Musajo, L., Dall'Acqua, F., and Caporale, G.,** La fotoreazione tra psoralene e flavin-monocleotide, *Gazz. Chim. Ital.,* 94, 1073, 1964.

266. **Logani, M. K., Austin, W. A., Shah, B., and Davies, R. E.,** Photooxidation of 8-methoxypsoralen with singlet oxygen, *Photochem. Photobiol.,* 35, 569, 1982.

267. **Kepka, A. G. and Grossweiner, L. I.,** Flash photolysis and inactivation of aqueous lysozyme, *Photochem. Photobiol.,* 13, 195, 1973.

268. **Matheson, I. B. C., Etheridge, R. D., Kratowich, N. R., and Lee, J.,** The quenching of singlet oxygen by amino acids and proteins, *Photochem. Photobiol.,* 21, 165, 1975.

269. **Cannistraro, S. and Van de Vorst, A.,** ESR and optical absorption evidence for free radical involvement in the photosensitizing action of furocoumarin derivatives and for their singlet oxygen production, *Biochim. Biophys. Acta,* 476, 166, 1977.

270. **Foote, C. S., Shook, F. C., and Abakerli, R. A.,** Chemistry of superoxide ion. IV. Singlet oxygen is not a major product of dismutation, *J. Am. Chem. Soc.,* 102, 2503, 1980.

271. **De Mol, N. J. and Beijersbergen van Henegouwen, G. M.,** Formation of singlet molecular oxygen by 8-methoxypsoralen, *Photochem. Photobiol.,* 30, 331, 1979.

272. **Foote, C. S. and Denny, R. W.,** Chemistry of singlet oxygen. XIII. Solvent effects on the reaction with olefins, *J. Am. Chem. Soc.,* 93, 5168, 1971.

273. **Chrysochoos, J. and Grossweiner, L. K.,** Flash photolysis of several aromatics by ultraviolet light and visible light in the presence of eosin Y, *Photochem. Photobiol.,* 8, 193, 1968.

274. **Kepka, A. and Grossweiner, L. I.,** Photodynamic oxidation of iodide ion and aromatic amino acids by eosin, *Photochem. Photobiol.,* 14, 621, 1972.

275. **Kasche, V. and Lindquist L.,** Transient species in the photochemistry of eosin, *Photochem. Photobiol.,* 4, 923, 1965.

276. **Nemoto, M., Usui, Y., and Koizumi, M.,** *Bull. Chem. Soc. Jpn.,* 40, 1035, 1967.

277. **De Mol, N. J. and Beijersbergen van Henegouwen, G. M.,** Relation between some photobiological properties of furocoumarins and their extent of singlet oxygen production, *Photochem. Photobiol.,* 33, 815, 1981.

278. **Kraljic, I. and Mohsni, E.,** A new method for the detection of singlet oxygen in aqueous solution, *Photochem. Photobiol.,* 28, 577, 1978.

279. **Rosenthal, I. and Pitts, J. N.,** Reactivity of purine and pyrimidine bases toward singlet oxygen, *Biophys. J.,* 11, 963, 1971..

280. **Korycka-Dahl, M. and Richardson, T.,** Photogeneration of superoxide anion in serum of bovine milk and in model systems containing riboflavin and amino acids, *J. Dairy Sci.,* 61, 400, 1977.

281. **Hurst, J. D., McDonald, J. D., and Schuster, G. B.,** Lifetime of singlet oxygen in solution directly determined by laser spectroscopy, *J. Am. Chem. Soc.,* 104, 2065, 1982.

282. **Parker, J. D. and Stanbro, W. D.,** Optical determination of the collisional lifetime of singlet molecular oxygen [O_2 ($^1\Delta_g$)] in acetone and deuterated acetone, *J. Am. Chem. Soc.,* 104, 2067, 1982.

283. **Ogilby, P. R. and Foote, C. S.,** Chemistry of singlet oxygen. XXXVI. Singlet molecular oxygen ($^1\Delta_g$) luminescence in solution following pulsed laser excitation. Solvent deuterium isotope effects on the lifetime of singlet oxygen, *J. Am. Chem. Soc.,* 104, 2069, 1982.

284. **Rodgers, M. A. J. and Snowden, P. T.,** Lifetime of O_2 ($^1\Delta_g$) in liquid water as determined by time-resolved infrared luminescence measurements, *J. Am. Chem. Soc.,* 104, 5541, 1982.

285. **Gorman, A. A., Hamblett, I., and Rodgers, M. A. J.,** Time-resolved luminescence measurements of triplet-sensitized singlet-oxygen production: variation in energy-transfer efficiencies, *J. Am. Chem. Soc.,* 106, 4679, 1984.

286. **Ronfard-Haret, J. C., Averbeck, D., Bensasson, R. V., Bisagni, E., and Land, E. J.,** Some properties of the triplet excited state of the photosensitizing furocoumarin: 3-carbethoxypsoralen, *Photochem. Photobiol.,* 35, 479, 1982.

287. **Sherman, W. V. and Grossweiner, L. I.,** Photobinding of psoralen and 8-methoxypsoralen to calf thymus DNA, *Photochem. Photobiol.,* 34, 579, 1981.

288. **Goyal G. C. and Grossweiner, L. I.,** The effect of DNA binding on initial 8-methoxypsoralen photochemistry, *Photochem. Photobiol.,* 29, 847, 1979.

289. **De Mol, N. J., Beijersbergen van Henegouwen, G. M., and van Beele, B.,** Singlet oxygen formation by sensitization of furocoumarins complexed with, or bound covalently to DNA, *Photochem. Photobiol.,* 34, 661, 1981.

290. **Salet, C., De Sa, E., Melo, T. M., Bensasson, R. and Land, E. J.,** Photophysical properties of aminomethylpsoralen in presence and absence of DNA, *Biochim. Biophys. Acta,* 607, 379, 1980.

291. **Anders, A., Poppe, W., Herkt-Maetzky, C., Niemann, E. G., and Hofer, E.,** Investigations on the mechanism of photodynamic action of different psoralens with DNA, *Biophys. Struct. Mech.,* 10(1—2), 11, 1983.

292. **Song, P. S., Chin, C. A., Yamazaki, I., and Baba, H.,** *Int. J., Quantum Chem. Quantum Biol. Symp.,* 2, 1, 1975.

293. **Beaumont, P. C., Parsons, B. J., Navaratnam, S., and Phillips, G. O.,** A laser flash photolysis and fluorescence study of aminomethyltrimethylpsoralen in the presence and absence of DNA, *Photobiochem. Photobiophys.,* 5, 359, 1983.

294. **Pathak, M. A., Allen, B. T., Ingram, D. J., and Fellman, J. H.,** Photosensitization and the effect of ultraviolet radiation on the production of unpaired electrons in the presence of furocoumarins (psoralens), *Biochim. Biophys. Acta,* 54, 506, 1961.

295. **Pathak, M. A.,** Mechanisms of psoralen photosensitization and *in vivo* biological action spectrum of 8-methoxypsoralen, *J. Invest. Dermatol.,* 37, 397, 1961.

296. **DeCuyper, J., Piette, J., and Van De Vorst, A.,** Activated oxygen species produced by photoexcited furocoumarin derivatives, *Arch. Int. Physiol. Biochim.,* 91(5), 471, 1983.

297. **Bensasson, R. V., Salet, C., Land, E. J., and Rushton, F. A. P.,** Triplet excited state of the 4′,5′ photoadduct of psoralen and thymine, *Photochem. Photobiol.,* 31, 129, 1980.

298. **Land, E. J. and Truscott, T. G.,** Triplet excited state of coumarin and 4′,5′-dihydropsoralen: reaction with nucleic acid bases and amino acids, *Photochem. Photobiol.,* 29, 861, 1979.

299. **Bensasson, R. V., Land, E. J., and Salet, C.,** Triplet excited state of furocoumarins: reaction with nucleic acid bases and amino acids, *Photochem. Photobiol.,* 27(3), 273, 1978.

300. **Bensasson, R. V., Chalvet, O., Land, E. J., and Ronfard-Haret, J. C.**, Triplet, radical anion and radical cation spectra of furocoumarins, *Photochem. Photobiol.*, 39, 287, 1984.

301. **Melo, M. T., Sa, E., Averbeck, D., Bensasson, R. V., Land, E. J., and Salet, C.**, Some furocoumarins and analogs: comparison of triplet properties in solution with photobiological activities in yeast, *Photochem. Photobiol.*, 30, 645, 1979.

302. **Peak, U. J., Peak J. G., Foote, C. S., and Krinsky, N. I.**, Oxygen-independent direct deoxyribonucleic acid backbone breakage caused by rose bengal and visible light, *J. Photochem.*, 25, 309, 1984.

303. **Pawelek, M. S.**, Factors regulating growth and pigmentation of melanoma cells, *J. Invest. Dermatol.*, 66, 201, 1976.

304. **Craw, M., Bensasson, R. V., Ronfard-Haret, J. C., Melo, M. T., Sa, E., and Truscott, T. G.**, Some photophysical properties of 3-carbethoxypsoralen, 8-methoxypsoralen and 5-methoxypsoralen triplet states, *Photochem. Photobiol.*, 37, 611, 1983.

305. **Krinsky, N. I. and Deneke, S. M.**, Interaction of oxygen and oxy-radicals with carotenoids, *J. Natl. Cancer Inst.*, 69, 205, 1982.

306. **Giles, A., Jr., Wamer, W., and Kornhauser, A.**, In-vivo protective effect of beta carotene against psoralen phototoxicity, *Photochem. Photobiol.*, 41(6), 661, 1985.

307. **Potapenko, A. Y., Sukhorukov, V. L., and Davidov, B. V.**, 8-methoxypsoralen sensitized photooxidation of tocopherols, *Photobiochem. Photobiophys.*, 5(2), 113, 1983.

308. **Potapenko, A. Y., Abiev, G. A., and Pliquett, F.**, Inhibition of erythema of the skin photosensitized with 8-methoxypsoralen by *alpha* -tocopherol, *Bull. Exp. Biol. Med. (U.S.S.R.)*, 89, 611, 1980.

309. **Craw, M. and Lambert, C.**, The characterisation of the triplet state of crocetin, a water soluble carotenoid, by nanosecond laser flash photolyses, *Photochem. Photobiol.*, 38, 241, 1983.

310. **Averbeck, D. S., Averbeck, F., and Dall'Acqua, F.**, Mutagenic activity of three monofunctional and three difunctional furocoumarins in yeast *(Saccharomyces cerevisiae)*, *Farmaco Ed. Sci.*, 36, 492, 1981.

311. **Averbeck, D.**, Photobiology of furocoumarins, in *Trends in Photobiology*, Helene, C., Charlier, M., Montenay-Garestier, T., and Laustriat, G., Eds., Plenum Press, New York, 1982, 295.

312. **Fujita, H. and Kitakami, M.**, Effect of paramagnetic species on 8-methoxypsoralen-sensitized photo-inactivation of bacteriophage lambda, *Photochem. Photobiol.*, 26, 647, 1977.

313. **Laskin, J. D., Lee, E., Laskin, D. L., and Gallo, M. A.**, Psoralens potentiate ultraviolet light-induced inhibition of epidermal growth factor binding, *Proc. Natl. Acad. Sci. U.S.A.*, 83, 8211, 1986.

Chapter 5

PSORALENS AS PROBES OF NUCLEIC ACID STRUCTURE AND FUNCTION

Paul L. Wollenzien

TABLE OF CONTENTS

I. INTRODUCTION

Psoralen photochemical cross-linking has a unique role among the current approaches for the elucidation of nucleic acid structure because of several desirable aspects of the chemistry and photochemistry of the process. The parent compounds 8-methoxypsoralen (8-MOP), 5-methoxypsoralen (5-MOP), psoralen, and the bent psoralen, angelicin, are planar heterocyclic compounds that show strong dark-binding affinities to nucleic acids.[1,2] These strong inter-actions occur because these compounds are able to intercalate between adjacent base pairs of double-stranded DNA,[3-5] DNA-RNA duplexes,[6] and the double-stranded regions within the folded back structures of single-stranded DNA and RNA.[7] New derivatives designed for their high solubility[8] and for their side chains that have special reactivities[9-11] also exhibit these high binding affinities to nucleic acids. This gives the psoralens a high specificity for nucleic acids even when they are reacted with nucleoprotein or are used for in vivo reactions.

Second, the linear psoralens possess geometrical structures such that when they are in an intercalated complex, double bonds in the furan and pyrone part of their structures are placed contiguous to the double bonds of pyrimidine bases of the complementary antiparallel strands; when activated, each of these sites can form covalent attachment to the adjacent base. If opposite adjacent bases are both pyrimidines, a cross-link between the two strands can be formed.[12,13] These monoadducts and cross-links give the photochemical reaction a unique signature. There are a variety of physical, chemical, and biological methods for detecting the presence and location of the psoralen adduct. When such information is available, it is possible to conclude that the intercalation site was accessible to psoralen binding and, in cases in which the secondary structure of the molecule is not known, to propose one.

Third, the covalent reactions of psoralens with nucleic acids are photochemical processes that can be activated with light in the wavelength range from 320 nm to 400 nm. These wavelengths can be produced in the laboratory with irradiation devices such that the desired reaction can be very selectively activated. This allows for convenience in the use of psoralens, and since these wavelengths are not absorbed very much by other biological components of the cell, very little photodamage is done as undesirable side reactions. In addition, the reaction of psoralen with the nucleic acids has been an active area of investigation for many years, and the photochemical process and the products of the reactions have been well characterized. The photoreaction is a two-photon process, and the steps have slightly different action spectra.[2,14] This allows the two processes to be separated; the nucleic acid can be manipulated to reveal alternate conformational forms, and in vitro reactions can be done to reassemble the molecule into its functional state before the second part of the photoreaction is completed. Once formed, the photochemical cross-links can be reversed by short-wave-length UV light to allow the separation and characterization of the nucleic acid site. These properties make the psoralen reagents valuable structural tools for the investigation of many interesting problems.

This chapter will deal with the current uses of psoralens in the elucidation of both DNA and RNA structure and function. It will be mainly concerned with those methods and approaches that give specific information about the arrangement of the molecules being investigated. Details of psoralen-DNA photochemistry, a description of the synthesis of new psoralen reagents, and the biological effects of psoralen in the cell are addressed in other chapters of this volume. Additional topics have been addressed in recent review articles,[1,2,14-16] and the reader will be directed to those for further information.

II. REACTION OF PSORALENS WITH DEOXYRIBONUCLEIC ACIDS

The psoralens were originally characterized as a class of naturally occurring compounds and later were recognized as the pharmacological agents in folk medicines that alleviated the symptoms of vitiligo and psoriasis through photochemical action.[1] Only recently was it established that their mode of action was through covalent interaction with DNA. Reactions with DNA through either the furan moiety as the 4′,5′ furan-side monoadduct or through the pyrone moiety as the 3,4 pyrone-side monoadduct were thought to be the principal modes of attachment.[17-21] Two groups made the observation in the same year (1970) that the reaction could occur such that both reactive sites were utilized to produce a cross-link between the two strands of the DNA duplex.[22,23] This provided the initial hypothesis for the killing properties of psoralen photochemical reaction on cells such that the inability of the cell to replicate its cross-linked DNA caused it to die.[24] In addition, the reaction of psoralen with ribonucleic acids was demonstrated at about the same time. However, this reaction and the reaction with single-stranded DNA occur to a much lower extent overall,[25] and these reactions have not been investigated with the same interest until recently.

A. How Reaction with Bulk Chromatin Reveals the Nucleosome Pattern

The basis for many experiments in which psoralen is used as a probe for chromatin structure is the ability of the reagent to penetrate cells or isolated nuclei, bind to accessible DNA, and upon irradiation form covalent cross-links. Moreover, a distinct advantage of this method is that the native structure can be probed with little disturbance of the nuclear organization and the sample is not destroyed as is the case with nuclease probes.

Hanson et al.[26] treated nuclei isolated from *Drosophila* embryos in this manner and then observed the pattern of cross-linking in the DNA by denaturing it with formaldehyde, spreading the DNA by the monolayer technique and viewing it in the electron microscope. This sample was compared to a control sample (purified DNA) that was cross-linked and prepared for electron microscopy in the same way. The chromatin sample contained series of open loops that formed a periodic distribution of sizes with the principal mode having a size of 160 to 200 nucleotides. About 60% of the DNA was in these open loops. Under the same conditions, the control DNA was very highly cross-linked and looked double stranded in the electron microscope. The size of the regions protected from psoralen cross-linking in chromatin is very close to the size of the repeat unit of the nucleosome, and this suggested that the pattern results from cross-linking in the linker region only.

This was further investigated by two groups who took advantage of the ability of micrococcal nuclease to digest selectively the linker region.[27,28] After chromatin was cross-linked with radioactive psoralen to a saturating level of about 20 to 27 trimethylpsoralen per 1000 base paris (bp), it was challenged with micrococcal nuclease. The presence of the psoralen did not interfere with the normal digestion pattern,, and the radioactive psoralen was lost from the samples at early times of digestion. This indicated that the primary location of the psoralen was in the linker.[27,28] The location of the psoralen cross-links in the linker region was confirmed by examination of the dinucleosomes and trinucleosomes in the electron microscope[27] and by fluorography of the samples after electrophoresis after they had been digested to different extents with micrococcal nuclease.[28]

These results together establish that psoralen photochemical cross-linking and electron microscopy as an assay is able to give a record of the organization of the DNA in the chromatin with 200-bp open loops being indicative of the nucleosome structure from many different sources. This pattern has been established for the chromatin of mouse liver DNA,[29] for the chromatin of SV40 DNA both as intracellular particles and as chromatin isolated from virus particles,[30] and for the chromatin of DNA isolated from *Drosophila* nuclei, both for the main band DNA and for three of the satellite DNA classes.[31]

Additional controls for this type of experiment have recently been reported by Conconi et al.,[32] who investigated the extent to which psoralen and UV treatment causes protein dissociation and structural alterations in solubilized rat liver chromatin. This is important because it is known that other reagents that bind by intercalation can cause nucleosomes to slide under some conditions.[33] They incubated psoralen-treated chromatin and control chromatin with increasing concentrations of sodium chloride and then fractionated the samples on sucrose gradients and determined that, in fact, the dissociation of the histones was similar in the two samples. In addition, to demonstrate that psoralen cross-linking is able to detect structural changes, they reacted H1-depleted chromatin at pH 10 with psoralen. Under these conditions, chromatin is unraveled in a fully reversible manner.[34,35] When the DNA was extracted from this sample and spread for electron microscopy under denaturing conditions, no single-stranded bubbles characteristic of the nucleosome core structure were seen. Instead, the DNA was highly cross-linked, indicating accessibility of the drug under these conditions.[32]

Psoralen is an intercalative reagent that unwinds the DNA helix, reducing the negative superhelicity of the DNA molecule when it binds.[4,5,36] Psoralen derivatives have higher binding constants to negatively supercoiled DNA than to relaxed DNA,[5,36] and covalent photoreaction gives a stable record of this. Sinden et al.[37] have taken advantage of this fact to measure the degree of supercoiling in the DNA molecule of interest in vivo. Circular duplex DNA isolated from all organisms has a natural negative superhelicity. In *Escherichia coli*, trimethylpsoralen interacts with the DNA as it does for purified supercoiled DNA molecules; if the DNA is relaxed by gamma irradiation or the drug coumermycin, which inhibits the action of DNA gyrase, the amount of trimethylpsoralen decreases. This indicates that the supercoiling of the DNA in vivo is unrestrained. On the other hand, if the same test is done for the DNA in HeLa cells or *Drosophila* cells, superhelicity in the DNA is not detected. This indicates that, although purified DNA from eukaryotes is supercoiled, within the limits of detection, all of the superhelical tension is restrained in nucleosomes or other nucleosome-like structures.[37]

Mitra et al.[38] have taken advantage of the fact that psoralen binds principally with the linker DNA of the nucleosome to investigate the orientation of the linker with respect to the chromatin superhelical axis oriented by the electrical field. They have done this by measuring the photochemical dichroism of psoralen addition and the optical dichroism of psoralen in the oriented sample. From these dichroism values, they were able to calculate the orientation of the nucleosome disks (33°) and the orientation of the linker region for two possible models: the linker as a semicircular arc and the linker as a straight rod. The values for these two models are an inclination of 7° or 45° with respect to superhelical axis. Because the linker is of variable length, a definite model cannot be proposed, but some other models can be ruled out.

B. Reaction of Psoralen with Altered Chromatin Structure

Since the previous studies demonstrated that DNA in nucleosomes in bulk chromatin is preferentially protected from psoralen photoaddition, psoralen derivatives can be used to determine if the normal nucleosomal repeat pattern is always present. Two general questions have been addressed by this method: whether specific sections of DNA, particularly transcriptionally active DNA or DNA being repaired, contain nucleosomes at all and whether there are sometimes alternative varieties of the nucleosome particle in some types of chromatin.

The small size of SV40 DNA allowed Hallick and co-workers to investigate in detail the distribution of the radioactive psoralen adducts after reaction of intracellular SV40 chromatin[39] or SV40 virions.[40] After photoreaction under these conditions, the SV40 DNA was purified and digested with restriction nucleases; the fragments were separated by electrophoresis, and the radioactivity in each was determined. The incorporation was compared to the distribution of radioactivity in SV40 DNA reacted with psoralen to the same extent. For intracellular SV40 chromatin, a 400-bp fragment was preferentially labeled compared to the DNA sample.[39] This fragment encompasses the origin of DNA replication,[41,42] the T antigen binding sites,[43-46] and the transcription start sites.[47] Previous electron microscopy experiments[48,49] and nuclease experiments[50-54] had indicated a nucleosome-free region at this location. For the SV40 DNA that had been treated in virions, however, no enhanced psoralen binding in this particular region was found. In addition, the amounts of psoralen addition to the DNA in the virion at saturation (175 to 210 hydroxymethyltrimethylpsoralens per genome) was significantly higher than could be achieved for the DNA in intracellular chromatin (105 hydroxymethyltrimethylpsoralens per genome).[40] This result is consistent with the observation that, although the virion contains all four histones, they may not be assembled into the normal nucleosome structure.[55]

The question of the distribution of nucleosomes in SV40 chromatin has also been addressed by Carlson et al.[56] using the fact that the presence of a nearby psoralen adduct in the DNA inhibits the double-strand activity of Hind III restriction nuclease. Therefore, if a particular Hind III site in the SV40 genome is preferentially uncut, at increasing levels of total psoralen addition, it would indicate that site is not covered by a nucleosome. The intrinsic rate constants at which Hind III sites in DNA are inactivated vary, apparently by influence of neighboring sequences, so this has to be taken into account. For the Hind III sites that were analyzed, there was no preferential protection, indicating random or near random distribution of the nucleosomes. However, the Hind III site near the origin of replication was not specifically studied.[56]

There is general agreement that mitochondrial DNA (mtDNA) does not possess a typical nucleosomal structure;[57] however, physical evidence has suggested the presence of a few distinctive proteins associated with mtDNA.[58,59] Two groups have investigated this possibility by psoralen cross-linking experiments. Potter et al.[60] and Pardue et al.[61] examined the cross-linking pattern by electron microscoy in *Drosophila melanogaster* and *Drosophila virilis* mtDNA. In both cases, in a large fraction of the molecules (>75%), there is protection from cross-linking in the A + T-rich regions that contain the origin of replication. In *D. melanogaster*, there is a series of five distinct protected regions; four were 400 bp long, and the fifth was 200 bp long. These are arranged upstream from the direction of replication.[60] In *D. virilis*, the A + T region is smaller,[62] and the protected region consists of two blocks of about 400 bp each and a smaller region of about 200 bp.[61] The orientation of the protected region is over and slightly downstream from the origin of replication in *D. virilis*. De-Francesco and Attardi[63] have cross-linked HeLa mtDNA both in whole cells and in isolated mitochondria, and analyzed the distribution of cross-links by electron microscopy. In at least 55% of the molecules the region around the origin is protected from psoralen cross-linking by proteins or protein complexes associated with the DNA. The protected segment is a single large region 300 to 1500 bp long and overlaps extensively the D loop region at the origin of replication; it was ruled out that for most of the molecules this appearance was not due to the presence of the D loop.[63] The large size of these protected regions and the location with respect to the origin of replication argues against the protection being due to proteins that function directly in DNA replication. Rather, these regions might have a structural role in anchoring the mtDNA to the inner mitochondrial membrane.[61]

The extrachromosomal ribosomal DNA containing the ribosomal RNA transcriptional units and the nontranscribed spacer regions have been investigated as models for the organization of active vs. inactive chromatin. Cech and Karrer[64] analyzed the distribution of trimethylpsoralen cross-links in *Tetrahymena thermophila* rDNA by quantitation of the looped pattern of the DNA in the electron microscope. They concluded that the spacer regions were protected from psoralen in vivo while the gene regions were relatively accessible. The distribution of the distance between cross-links in the transcribed regions showed a pattern of 215, 335, and 460 bp; this is consistent with an altered conformation of nucleosomes or nucleoprotein structure containing proteins other than histones. Only in the terminal spacer regions did the cross-linking pattern indicate a regular nucleosome structure.[64] Sogo et al.[65] treated nucleoli or nuclei from *Dictyostelium discoideum* with trimethylpsoralen and analyzed the distribution of psoralen by electron microscopy. At the high levels of psoralen addition they used, they saw a very striking difference between the amount of cross-linking in the transcribed region, which looked nearly completely double stranded, and that in the non-transcribed regions, which exhibited the open looped pattern characteristic of the nucleosome structure. In addition, the high concentration of psoralen in the transcribed regions was confirmed by an electrophoresis assay in which fragments generated by restriction endonuclease digestion have reduced mobility according to how much psoralen they contain.[65]

In their characterization of an immunosuppressive variant of minute virus of mice (MVMi), Doerig et al.[66] identified a 40S nucleoprotein complex that contains the double-stranded replicative form of the virus. They investigated the organization of this particle by staphylococcal nuclease digestion and did not observe a pattern characteristic of nucleosome structure. To investigate this further, a nuclear extract containing the MVMi nucleoprotein particles was psoralen-cross-linked, and the DNA was purified and examined in the electron microscope. The sizes of the denatured open regions in the replicative-form DNA followed a fairly narrow distribution with a mean length of 90 nucleotides. An SV40 DNA control sample prepared in the same way showed open regions with a mean length of 160 nucleotides, characteristic of its known nucleosomal structure. MVMi RF DNA is known to be transcriptionally and replicatively active, and in addition it is a linear molecule, so it cannot be supercoiled; either or both of these facts could account for its lack of normal nucleosome structure.[66] Investigations on the structure of herpes simplex virus type 1 (HSV-1) using micrococcal nuclease digestion indicated that the replicated DNA was very sensitive compared to host DNA.[67] Whether newly replicated HSV-1 DNA was assembled into nucleosomes and was supercoiled was determined by a quantitative analysis of trimethylpsoralen incorporation by Sinden et al.[68] At 4 hr after infection after the initiation of DNA replication, the incorporation of psoralen into HSV-1 increased four times relative to the incorporation into the host DNA. This increase was not affected by gamma irradiation that produces single-strand breaks in the HSV-1 DNA. This suggests that most of the replicating HSV-1 DNA is free of torsional stress and that the differences in incorporation are due to accessibility of the DNA by the psoralen.[68]

Psoralen reaction with chromatin after DNA damage has been used to determine the distribution and movement of nucleosomes during the repair processes. Cleaver[69] has used psoralen addition to living cells in order to characterize the stability of chromatin during normal growth and during DNA repair and replication. In order to ensure the survival of both normal and repair-deficient cells, very low levels of psoralen addition were used (less than one psoralen per 100 nucleosomes). He exploited the known nuclease sensitivity of psoralen in linker regions[27,28] to determine the amount of psoralen in the linker and in the nucleosome. The distribution under these very low levels of addition is about equal amounts of psoralen in the two regions; this represents only a two- to threefold higher affinity of psoralen per nucleotide to the linker region when the relative amount of DNA in the linker and nucleosome are taken into account.[69] When the cells were allowed to grow after exposure to psoralen, after 15 hr the distribution of the psoralen became randomized, indicating a slow movement of the nucleosomes. A more rapid movement of the nucleosomes occurs for about 1 hr in those regions of repair replication, as demonstrated by the nuclease sensitivity of repair patches after treatment with 5-methylisopsoralen and UV light.

An experiment similar to this was reported by Mathis and Althaus.[70] Pyrimidine dimers were formed with 254-nm light, or chemical adducts were formed with *N*-acetoxy-2-acetylaminofluorene in cultured hepatocytes, and the cells were allowed to repair this damage for different times before 8-methoxypsoralen and 365-nm light were applied. The amount of incorporated psoralen compared to control samples prepared in parallel was used to estimate the fraction of accessible DNA. Under conditions where 80% of the cells underwent repair synthesis stimulated by 254-nm light, there were two waves of DNA accessibility at 20 min and at 3 hr. When the chemical adduct was used to induce repair synthesis, there was a corresponding decrease in the DNA accessibility at 20 min.[70]

C. Stabilization of Transient Structures in DNA, Chromatin, and DNA-RNA Duplexes

A number of different types of structures in nucleic acids are inherently unstable when purified due to the unfavorable thermodynamic situation that a completely base-paired duplex will be more stable than a structure that has some unpaired nucleotides. These are palindromic sequences in double-stranded DNA in the form of cruciforms, D loop stuctures in which an RNA single strand invades a DNA duplex, R loop structures in which an RNA single strand invades a DNA duplex, transcription complexes, and replication complexes. Psoralen photochemical cross-linking is able to stabilize these structures so that they can be purified and characterized. This allows new experimental approaches to determine the importance of these structures in the cell.

An early attempt to identify cruciforms in mouse DNA was made by Cech and Pardue.[71] Trimethylpsoralen and UV light were used to cross-link the DNA in mouse tissue culture cells, and the DNA was purified and examined in the electron microscope. Based on the known frequency of inverted repeated sequences in the mouse genome, they estimated that the 2000 molecules they scanned contained 1000 inverted repeats that were potentially observable. Since only three possible cruciforms were seen, they concluded the absence of cruciforms in the sample. *Tetrahymena* rDNA which contains an inverted repeat, treated in vitro with psoralen after denaturation and quenching, acted as the positive control because in fact in those molecules cruciforms could be seen.[71] More recently Cech and Karrer[64] reexamined the chromatin structure of *T. thermophila* rDNA; again no cruciforms were found.

Sinden et al.[72] exploited the 76-nucleotide inverted repeat sequence at the *lac* operator to study the occurrence of a cruciform. This sequence contains two Eco RI restriction nuclease sites, so that if the cross-linked linear form is digested a 66-nucleotide fragment is released, but if the cross-linked cruciform is digested two 33-nucleotide fragments are released. A restriction fragment containing this sequence was moved onto a plasmid, and the frequency of the cruciform was determined at different degrees of supercoiling. The DNA was psoralen cross-linked, digested with Eco RI, and then analyzed by gel electrophoresis. The cruciform exists at a sufficiently high superhelical density (< -0.04) but does not exist in the cell.[72] This assay has also been used to study the effects of temperature and ionic strength on the rates of the transition between the linear and cruciform forms.[73]

Recently Hsu has reported electron microscopic evidence for two cruciform structures in intracellular SV40 DNA.[74] The approximate location of these forms in the SV40 genome was determined; one is in the region of the origin of replication. These molecules occur at very low frequency (0.1 to 0.2%), and the precise sequence has not yet been determined. Caffarelli et al.[75] have studied the persistence of cruciform structures on supercoiled pBR322 and ColE1 DNA, and used trimethylpsoralen photoreaction to show that the preferential location for nucleosomes on pBR322, when the histone-DNA ratio was 0.5, was on the A + T-rich portion of the DNA.

The cross-linking of RNA in a transcription complex to its template (*E. coli* RNA polymerase transcribing SV40 DNA) was first demonstrated by Shen and Hearst.[76] This technique has been used as part of the characterization of the eukaryotic ternary transcription complex by Sargan and Butterworth.[77,78] In their method, endogenous nuclease activity in nuclei releases ternary transcription complexes; these are collected and are labeled by in vitro run on transcription. The radioactive RNA is cross-linked to the DNA and then trimmed with RNase A before further analysis. The size of the DNA in this complex[77] and secondary nuclease digestion of the complex indicate a digestion barrier characteristic of nucleosomal histones.[78] This suggests that the transcription complexes are in close proximity to histone-containing, nucleosome-like particles.[78]

Two studies have used in vivo psoralen cross-linking as a method of fixing nucleic acid secondary structure in replicating adenovirus[79] and in the replication intermediate of poliovirus.[80] This is important since in both of these systems there has been considerable question about rearrangement of the strands during purification. Revet and Benichou[79] fixed replicating adenovirus DNA to prevent branch migration and then were able to observe by electron microscopy all branch positions characteristic for the first round of replication. They concluded that the growing points progressed from either end of the molecule, but for a given replicating molecule, DNA synthesis is initiated at one end and not at both ends.[79] The replicative intermediate of poliovirus contains multiple nascent RNA molecules replicating off of the poliovirus genome which is a single-stranded RNA chain of positive polarity. The question whether the backbone of the replicating structure was predominantly single stranded or double stranded could not be resolved because of considerable uncertainty about whether the complementary strands were renaturing during purification.[81] Richards et al.[80] psoralen cross-linked the replicative intermediate in vivo and then examined the structures by electron microscopy after denaturation; they observed only single strands of heterogeneous length, indicating that the replicative intermediate in the cell contains little or no duplex structure. Under the same conditions, replicative form poliovirus, which is known to be double stranded in the cell, was extensively cross-linked.[80]

Psoralen cross-linking has been used to fix RNA-DNA duplexes to ensure their stability during subsequent experimental manipulations. Wittig and Wittig[82] formed tRNA-DNA R loops in which the tRNA was coupled to biotin. This R loop molecule was stabilized by cross-linking and then was purified by affinity chromatography on an avidin-glass column. The purified double-stranded DNA could be used for cloning. Chatterjee and Cantor[83] studied the repair in *E. coli* of a plasmid molecule contained an inserted R loop. They formed aminomethyltrimethylpsoralen monoadducts (with 390-nm light) onto 16S ribosomal RNA, formed R loop structures with a plasmid containing part of the rDNA, and then fixed the duplex by reirradiation with 365-nm light. In this way the psoralen adducts were confined to the R loop region only. The fixed duplex (with or without nuclease S1 treatment) was then transformed into *E. coli*. Several types of large deletions in the plasmid resulted from this treatment.[83]

RNA-DNA duplexes have been fixed for electron microscopy utilizing psoralen cross-linking by two groups. Kaback et al.[84] lightly psoralen cross-linked the DNA molecule before R loop formation to prevent complete strand dissociation during formation of the R loop, and then the complex was treated with glyoxal to prevent rearrangement. Wollenzien and Cantor[85] covalently attached DNA restriction fragments to RNA molecules at selected locations in order to identify the polarity of the RNA in the electron microscope. This was done by forming aminomethyltrimethylpsoralen monoadducts on the restriction fragments with 390-nm light, hybridizing the fragments to the RNA, and then reirradiating to form cross-links. After denaturation, the cross-linked DNA fragment was seen as a small cross-strand on the RNA molecule.[85]

A significant fraction of the mitochondrial DNA molecules from most animal cells contains a displacement loop (D loop) structure resulting from the synthesis of a 7S DNA which displaces the parental H strand.[86,87] The small infiltrated strand is lost when the mitochondrial DNA is linearized.[88] DeFrancesco and Attardi[63] have succeeded in cross-linking the D loop structure by psoralen-cross-linking HeLa cell DNA *in situ*. The frequency they observed for these structures corresponded to the frequency that had been measured in vivo.

D. Perturbation of the Structure of DNA upon the Reaction with Psoralen

The unwinding of the DNA helix upon psoralen binding[4,5] and addition[4] was mentioned in Section II.A. The crystal structure of an 8-MOP-thymine monoadduct has recently been determined.[89-91] The details of this structure provide additional information concerning changes in the structure of the DNA upon reaction with psoralen. The psoralen adduct is important because this monoadduct is the predominant monoadduct in DNA during the normal reactions with 365-nm light.[13] In the unit cell in the crystals obtained, there were three molecules of one handedness and three molecules of the opposite handedness. These correspond to the two stereoisomers that would result if psoralen forms a monoadduct in either a 5'...T-A...3' or 5'...A-T...3' dinucleotide site. For the three molecules of each handedness, there were different interplanar angles between the thymine and the psoralen (44 to 53°). If this structure is extrapolated to a possible structure for the cross-link, it indicates that a sharp kink (as much as 70°) may occur at the site where the psoralen is introduced.[89,90]

Sinden and Hagerman[92] have tested whether appreciable bends occur in DNA restriction fragments by two methods. In the first, psoralen-photoreacted molecules were examined by polyacrylamide gel electrophoresis. It is known that DNA can be stably bent as a result of the inherent conformation of particular sequences and this results in abnormal electrophoretic behavior on polyacrylamide gels.[93,94] Sinden and Hagerman digested pBR322 DNA with Hae III restriction nuclease, photoreacted the fragments with 8-MOP or trimethylpsoralen, and then electrophoresed the products on 5 or 9% polyacrylamide gels. The fragments were reduced in mobility, but not nearly as much expected if there were sharp bends in them. In the second test, the decay of birefringence following removal of an electric field that oriented the molecules was used to determine if the psoralen-modified DNA possessed a shorter relaxation time that would indicate a bent conformation. The DNA fragment studied contained an average (estimated) of eight adducts and four cross-links. No evidence of a shorter relaxation time was apparent; instead the data were consistent only with an extension of the linear length due to the intercalated psoralen. A DNA molecule having only one cross-link has not yet been studied by these methods.[92]

Psoralen photochemical addition has been used to study repair of DNA lesions in African green monkey cells by Zolan et al.[95] It was determined that the αDNA sequences were repaired less efficiently than the bulk of the genome. The chemical identity of hydroxy-methyltrimethylpsoralen adducts has been determined to find out if the αDNA contained some types of monoadducts that could not be repaired as well as the adducts present in the bulk DNA.[96] Both thymine-type stereoisomers were found in the αDNA and the bulk DNA at the same frequency, both in vitro and in vivo. However, the frequency of diadducts in the αDNA was markedly reduced for the sample that was reacted in vivo. This suggests a possible conformation constraint on the αDNA in the nucleosome that prevents the formation of psoralen cross-links.[95]

Transcriptional enhancers are DNA elements that are able to stimulate the expression of neighboring genes even when they are separated by large distances. Several proposals have been offered for the mechanism by which they operate. Two of the proposals are (1) the tracking of transcription factors from the site of the enhancer to the gene control region and (2) the alteration of chromatin structure by a factor that recognizes the enhancer sequence. For both of these mechanisms, the quality of the DNA in the interval between the enhancer and the gene would be important. Courey et al.[97] have tested these ideas by introducing psoralen-modified DNA fragments onto a plasmid that contains the SV40 enhancer and the gene for human β-globin. The plasmid was then transfected into HeLa cells, and the amount of transcription was measured. The psoralen-modified DNA did in fact reduce the amount of transcription if it was inserted between the enhancer and the gene; this effect was greatest when the inserted fragment was close to the transcription start site, but it was also significant when placed far away. Psoralen monoadducts had this effect. When the fragment contained 30 to 40 adducts but was 90% uncross-linked, transcription was reduced 60%. When the fragments were irradiated for a longer time and were 90% cross-linked, the amount of transcription did not decrease further. These results support either a scanning model or a structural model and rule out some other models for the action of the enhancer.[97]

E. Use of Psoralen Derivatives to Produce Interhelical Cross-Links and Nucleic Acid-Protein Cross-Links

The chemical structure of the psoralens is amenable to derivation, and this has been done to extend the versatility of the molecule. Cantor and co-workers[9,10] have synthesized a set of reagents for interhelical cross-linking, nucleic acid-protein cross-linking, and site-directed cross-linking. These have already been useful for a variety of problems. The arrangement of the DNA inside bacteriophage lambda was studied with a *bis* psoralen that is capable of producing cross-links between DNA duplexes.[98] After reaction in the bacteriophage, the DNA was purified, fragmented with a restriction nuclease, and examined by electron microscopy for the presence of cross-linked duplexes. Such molecules were seen; the frequency of the occurrence of different fragments indicated a complex but nonrandom arrangement of the DNA inside the bacteriophage. A psoralen that contained a succinimidyl group was also used to examine the distribution of DNA inside bacteriophage lambda.[10] In this experiment, the reagent was first reacted with the bacteriophage in the dark, and the excess psoralen was removed before photoreaction. The DNA was purified and fragmented with restriction nuclease, and the fragments were examined for rapid renaturation by a gel electrophoresis technique as an indication that they were cross-linked. All the fragments were partially cross-linked, but the frequency was different for different fragments. This again suggests a complex but nonrandom organization of the DNA in the bacteriophage. Welsh and Cantor[99] have recently reviewed these and other DNA-protein cross-linking methods.

A psoralen reagent that contains a sulfhydryl group has been used for site-directed mutagenesis.[9] This psoralen was delivered to a patch on a plasmid DNA that contained mercurated nucleotides. The excess psoralen was removed, the complex was irradiated to produce cross-links, and then the plasmid was used to transform an *E. coli* host that had been activated for repair. A pattern of repaired nucleotides was detected around the mutagenized site.[100]

Another set of psoralen reagents that are reactive with proteins has been recently described by Elsner et al.[11] These are methoxypsoralen-cystamine-arylazido reagents. The efficiency of these reagents has been determined by cross-linking DNA and histones in solubilized chromatin. The great advantage of these reagents is that the nucleic acid-protein cross-link can be easily reversed by reduction of the disulfide bond.

III. REACTION OF PSORALENS WITH SINGLE-STRANDED NUCLEIC ACIDS

The geometrical requirements for psoralen cross-linking between two bases (in polynucleotides) are fairly precise: the psoralen occupies an intercalated position between the bases; the bases must be separated by a distance that corresponds very closely to the hydrogen bonding distance.[12,13,89,90] Certainly, double-stranded DNA contains a high density of suitable intercalation and reaction sites, and it reacts very well with linear psoralens even though the efficiency of the process can be influenced by bulky side groups at some positions of the psoralen. The converse is true, at least to a first approximation: if psoralen cross-links two nucleotides in a single-stranded molecule, this indicates that the nucleotides were forming a base-paired interaction. Therefore, psoralen cross-linking can be used to obtain information on the hydrogen-bonded internal structure of the molecule. In most instances, these interactions are secondary structure; in other instances, the cross-linked site may belong in the category of hydrogen-bonded tertiary structure. The purpose of this section is to describe first the reaction of psoralens with single-stranded DNA molecules. Then, reactions of psoralen with tRNA and 5S RNA will be described as models for the reaction of psoralen with RNA molecules in general. Finally, the reaction of psoralen with 16S ribosomal RNA, precursor RNA, and complexes involving small nuclear RNA will be described.

A. Structure in Single-Stranded Deoxyribonucleic Acids Detected by Psoralen Cross-Linking

The early work of Shen and Hearst[101] and Shen et al.[102] demonstrated the usefulness of psoralen cross-linking in determining the folded-back regions in single-stranded DNA molecules. The structure of the origin of replication of fd bacteriophage was identified by performing cross-linking with 4,5',8-trimethylpsoralen and irradiation, and then examining the DNA in the electron microscope after denaturation. The cross-linking was performed at different salt (NaCl) concentrations to determine the most stable structures. One unique hairpin structure was visualized at low salt; the location of this feature was determined on single strands that were generated by restriction nuclease Hind III cleavage of the replicative form I of the phage and strand separation. The location of the first hairpin corresponded to a region of the DNA that is known to be important for the initiation of replication.[101] In addition, cross-linking experiments were also conducted on the fd DNA inside the virion. Most of the DNA molecules exhibited one single hairpin; mapping experiments determined that this was again a unique feature at the origin of replication.[102] This means that the fd DNA is situated in the virion in a specific nonrandom arrangement. Ikoku and Hearst[103] have examined the structure *in situ* of a chimeric phage of M13 and G4 in which a large section of G4, including its origin of replication, was inserted into M13 between the M13 most stable hairpin and the other stable hairpins that are present in the M13 origin of replication. The M13 most stable hairpin was the feature cross-linked in the virion. This again indicates a specific orientation of the phage in the virion. Both origins are functional; this means that the origin region need not be located at the end of the phage in order to be recognized for replication.[103]

The secondary structure of adenovirus DNA was studied by trimethylpsoralen cross-linking and electron microscopy by Peterlin et al.[104] to possibly correlate the sites of secondary structure with gene control regions. The rightward and leftward strands of the DNA were separated, cross-linked at different ionic strengths, and then examined. Orientation of the strands in the electron microscope was accomplished by heteroduplex formation with complementary strands that contain an asymmetric insert. The two strands have very different secondary structures; this is possibly significant because both strands contain the sequences present in RNA transcripts. However, no convincing match of secondary structure and gene structure could be made at the time.

B. Characterization of the Reaction of Psoralen with RNA in Transfer RNA and 5S RNA

Characterization of the products of the photochemical reaction between psoralen and ribonucleic acids indicates that uridine is the predominant target in a variety of substrates.[18,105-107] However, it is becoming apparent for a number of large natural RNA molecules that there are structural parameters that modulate the binding affinity of psoralen to different sites.[108-110] Transfer RNA and 5S RNA have been used as test molecules to characterize the specificity of psoralen addition and cross-linking, and also to test the methodology of the techniques that are used.

Bachellerie and Hearst[108] used large concentrations of [^{3}H] 4'-(hydroxymethyl)-4,5',8-trimethylpsoralen and low light doses to covalently bind the psoralen as monoadducts to *E. coli* tRNA^{Phe}. The molecule was fragmented at m^{7}G$_{46}$ by chemical cleavage and then further decomposed by ribonuclease T1 and secondary nuclease digestion. Uridine 51 was the major photoreactive site in the tRNA, binding about 50% of the hydroxymethyltrimethylpsoralen. All other uridines were reactive. The reactivity of the sites was also studied as a function of temperature and salt; different sites change differently during the structural transition, and Bachellerie and Hearst have rationalized some of these.[108] Garrett-Wheeler et al.[111] have studied the cross-linking pattern of hydroxymethyltrimethylpsoralen in yeast tRNA^{Phe}. They reacted the molecule in low salt with EDTA and multiple additions of the psoralen to obtain maximal levels of cross-linking (two psoralens per tRNA). The molecule was fragmented by a partial ribonuclease T1 digestion and end-labeled, and pairs of cross-linked oligonucleotides were separated by a two-dimensional gel electrophoresis technique in which the psoralen cross-links were mildly photoreversed before the second dimension electrophoresis. The oligonucleotides were purified and enzymatically sequenced to establish their identity. This technique was a modification of the procedure of Turner et al.[112] Garrett-Wheeler et al.[111] were able to identify five different cross-links in the tRNA. Every major stem except the anticodon arm was cross-linked, and in addition one tertiary structural feature was cross-linked. As well as finding one U-U cross-link (U_6-U_{68}), as would be expected, three C-U cross-links (U_8-C_{48}, U_{50}-C_{63}, and U_{52}-C_{63}) and one C-C cross-link (C_{11}-C_{25}) were found. Nielsen and Leick[113,114] have reacted 4'-aminomethyl-4,5',8-trimethylpsoralen and 8-MOP with yeast tRNA^{Phe}, and determined the sites of reaction of the latter. They found one major site at U_{50} (55% of the adducts) and two minor sites at U_{59} and C_{70}. Taken together, these studies point to the particular reactivity of the T stem with respect to psoralen binding and photoreaction. The determination of the cross-links in this stem[111] indicate that, in fact, the site is an intercalation site within the T stem; the tRNA must be able to accommodate the intercalated psoralen without too much disruption of its overall structure.

Rabin and Crothers[115] cross-linked [^{32}P]-labeled *E. coli* 5S RNA with aminomethyltrimethylpsoralen, fragmented the product with ribonuclease T1, purified the predominant new high-molecular-weight band, and photoreversed it to separate and purify the cross-linked oligonucleotides. These were subjected to secondary digestion for identification. The major cross-linked product yielded oligonucleotides that corresponded to the 5' terminal region (nucleotides 1 to 9) and a region (nucleotides 108 to 116) very close to the 3' terminus. In models of the 5S RNA constructed from sequence comparison,[116] these regions are base-paired. The site of the cross-link was not determined more closely. Thompson et al.[109] have reacted hydroxymethyltrimethylpsoralen with *Drosophila melanogaster* 5S RNA. The analysis was very similar to the one performed by Rabin and Crothers[115] with the addition that the base composition of the purified oligonucleotides was analyzed by paper electrophoresis at pH 3.5. These conditions permit the identification of the nucleotide-psoralen monoadduct.[108] Four pairs of cross-linked oligonucleotides were identified. For two of these the exact site of the cross-link could be inferred. The other two were proposed based on possible complementarity within the molecule. Two U-U and two U-C cross-links were found. Interestingly, all of the sites occurred at a UpU site at the end of a helix or in a weak helix. This finding suggests that psoralen is reacting preferentially at sites where there can be local weakness in the helical structure.[109,110] The secondary structure predicted for the *D. melanogaster* 5S RNA from these results and base-pairing stability rules[117,118] is identical to the model for *Torulopsis utilis* 5S RNA[119] and is very similar to other models of 5S RNA.[116]

C. Use of Psoralens in the Study of Ribosomal RNA

For large RNA molecules, the possible complexity of the pattern of cross-linking will increase much faster than in direct proportion to the size of the molecule. This means that it is much more difficult to get a whole description of the cross-links in a large molecule. The initial experiments to determine the structure of ribosomal RNA by psoralen cross-linking were done on *D. melanogaster* 26S and 18S rRNA, and were analyzed by electron microscopy.[120] Purified rRNA was cross-linked with hydroxymethyltrimethylpsoralen at a series of salt concentrations and then examined for characteristic features. A cross-link in a large central hairpin of the 26S rRNA was observed, as was a cross-link in the 18S that allowed visualization of an open loop structure. This was a demonstration that interactions between residues distant in the primary sequence were important in the structure of the molecule.

The *E. coli* 16S ribosomal RNA has been the subject of extensive analysis because a great amount of structural information is available for it from a variety of sources, and this is of value in the interpretation of cross-linking results. Initially, electron microscopy was used after cross-linking the purified RNA or the RNA in the 30S ribosomal subunit.[121-123] Eleven open loop structures were mapped in the 16S rRNA cross-linked in solution,[121] and seven of these structures were cross-linked in the inactive 30S subunit.[122] The amount of psoralen incorporation into the 16S rRNA in the 30S subunit, particularly in the active form present in high magnesium buffer, was greatly reduced. Initially no correlation was found between these structures and the secondary structure predicted from sequence comparison.[124] The orientation of the structures cross-linked in the 16S rRNA in the inactive subunit[125] and in the purified RNA have been reexamined.[126] A total of 13 different types of structures were identified with a more sophisticated classification system. The orientation of part of these as originally reported was reversed; with the corrected orientation, seven of the cross-links correspond very well to base-paired interactions in the secondary structure. Not found in the proposed secondary structure are an additional set of six interactions cross-linked by psoralen; these cross-links, involving the 3'-terminal region of the molecule, are made both in the inactive 30S subunit and in the purified RNA.[126]

Chu et al.[127] have exploited psoralen monoaddition to determine whether these cross-links, the nonsecondary structure cross-links, are found in the active conformation of the 30S subunit. Psoralen monoadducts were made on the RNA with 390-nm light in the inactive form of the subunit in low magnesium; the magnesium was then increased to activate the subunits, and the cross-linking was performed. The frequency of cross-linking was determined by electron microscopy. There was a reduction in the frequency of one major class of cross-links in proportion to the fraction of the sample that was successfully reactivated. This clearly indicates a specific conformational change in the RNA during the activation process.[127]

The electron microscopy analysis provides general information about some of the cross-links that are present in the molecule, but the accuracy for identifying the location is on the order of 20 to 50 nucleotides.[126] Turner et al.[112] and Thompson and Hearst[128] have performed RNA-sequencing experiments in an effort to obtain the exact location of psoralen cross-links in the 16S rRNA. In the experiment of Turner et al.,[112] the cross-linked 16S rRNA was fragmented with ribonuclease T1, and the cross-linked nucleotides were identified by a two-dimensional gel electrophoresis method in which the cross-links were reversed before the second dimension. Cross-linked oligonucleotides would electrophorese as off-diagonal spots; they were purified and sequenced. One cross-link in a hairpin structure which is part of the secondary structure was found. In the experiment of Thompson and Hearst,[128] the cross-linked 16S RNA was fragmented, prefractionated through either a RPC-5 or a benzoylated naphtholated DEAE column to enrich for the cross-linked oligonucleotides, and then fractionated by two-dimensional gel electrophoresis. In their system, the first dimension was a 4% polyacrylamide gel, and the second was a 20% polyacrylamide gel run warm. Cross-linked oligonucleotides ran as off diagonal spots and were collected, photoreversed, separated, and sequenced for identification. Thirteen cross-links were identified. Seven of these cross-links were found in current versions of the secondary structure as it existed at the time, three of the cross-links helped to decide uncertain regions of the secondary structure, and three were new interactions definitely outside of the secondary structure.

These cross-links, particularly the ones indicating new contacts within the structure, enabled Thompson and Hearst[129] to propose a functional model for protein synthesis that involved a series of conformational changes within the 16S rRNA. This psoralen cross-linking information, along with chemical cross-linking information and other structural data, was used by Expert-Bezancon and Wollenzien to build a model of the arrangement for the 16S rRNA in the 30S subunit.[130]

A gel electrophoresis technique that involves the separation of intact 16S rRNA molecules based on the location of the cross-link they contain has been introduced by Wollenzien and Cantor.[125] In this method, molecules that contain long-distance cross-links have reduced mobility in a formamide polyacrylamide gel; because the molecules were cross-linked to low extent, this results in a large enrichment of adduct at the position of the particular cross-link. Hui and Cantor[131] have used a hybridization-nuclease challenge method to analyze molecules that were enriched in this way. By electron microscopy they contained a cross-link between positions at approximately 920 and 1500 to 1542. End-labeled DNA restriction fragments were hybridized to the RNA and then were digested with mung bean nuclease to cut the DNA at the position where it cannot hybridize properly to the cross-linking site. Based on their results, Hui and Cantor concluded that the cross-link was either (919-921)·(1539-1532) or (921-923)·(1532-1534). It has been possible to decide between these two choices, in favor of the latter, by performing reverse transcription on the same class of molecules using a synthetic DNA oligonucleotide as a primer.[132] This an extension of the method originally proposed by Youvan and Hearst.[110] This method works because reverse transcriptase is not able to polymerize the cDNA through the nucleotide that contains the cross-linking site. Reverse transcription stops at nucleotides 925 (the strongest), 924, 922, and 921 were detected. Unfortunately, the method cannot be used for the 3' site because of its closeness to the 3' terminus.[132]

D. Use of Psoralens in the Study of Messenger RNA, Precursor RNA, and snRNA

The secondary structure of RNA molecules can often be inferred by sequence comparison because of the proposition that common function derives from a common secondary structure.[116] The secondary structures for tRNA, 5S RNA, the ribosomal RNAs, and the core structure of some intervening sequences have been solved in this way. For messenger RNA and other classes of RNA, however, there is not sufficient sequence information available for this approach. Examples also exist for molecules that perform the same function yet have very different sequences, for example, M1 RNA from *E. coli* and Bacillus subtilis.[133] It is very important in these cases to have experimental verification of secondary structures.

Currey et al.[134] have analyzed the secondary structure of poliovirus RNA, a plus-strand RNA, by trimethylpsoralen cross-linking and electron microscopy after denaturation to determine the locations of hairpin and open loop structures (long-range interactions). This information was compared to the secondary structure predicted by the computer programs RNA2 and FOLD[135] that use a global free-energy calculation to identify the best secondary structure. Good correlation was found between the two structures; this gives credibility to the computer-aided method.

Pederson and co-workers have used psoralen cross-linking extensively in their analysis of the organization and function of heterogeneous nuclear RNA (hnRNA), a class of RNA that contains the precursors to cytoplasmic messenger RNA. It was known that hnRNA contained many double-stranded regions;[136,137] These are potentially important as processing sites; however, it was not known whether these double-stranded regions existed in vivo. Calvet and Pederson[138] treated HeLa cells with aminomethyltrimethylpsoralen and UV light. This treatment was able to cross-link a large fraction of double-stranded regions, as demonstrated by the property that the psoralen-treated, double-stranded RNA, subjected to a heat and quick cool step, retained its RNase resistance. This demonstrates the occurrence and accessibility of these regions in vivo.

These results have been extended and have established the in vivo association between small nuclear RNA (snRNA) and hnRNA.[139] HeLa cells were treated with aminomethyltrimethylpsoralen and UV light, and then high-molecular-weight RNA was isolated under denaturing conditions from nuclei or from subnuclear fractions. To reverse the cross-links and release small-molecular-weight species, 254-nm light was used. SnRNA U3 and 5.8S rRNA were released from the high-molecular-weight RNA isolated from the nucleolar subfraction. SnRNA U3 was identified on the basis of its mobility as compared to snRNA U3 isolated from nucleoli; 5.8S rRNA was identified on the basis of its mobility, and in addition it was eluted from the gel and fingerprinted to demonstrate its identity. SnRNA U1 and U2 were released from the high-molecular-weight RNA isolated from ribonucleoproteins which contain hnRNA. U2 was tentatively identified on the basis of its gel mobility; U1 was identified by hybridization to a cDNA clone specific to U1 RNA.[139] Subsequently, the identity of U2 was also confirmed by hybridization to a cDNA clone specific to U2 RNA.[140]

U1 RNA is complexed with nuclear proteins to form a ribonucleoprotein complex, U1 RNP.[141,142] Setyono and Pederson have demonstrated that it is the U1 RNP that is bound to hnRNA.[143] They did this by pulse-labeling HeLa cells to selectively label the pre-messenger RNA, isolating the nuclei and cross-linking them with aminomethyltrimethylpsoralen, and then purifying high-molecular-weight hnRNP particles on sucrose gradients made with 50% formamide. This material was applied to an immunoaffinity column which contained an antibody specific for U1 RNP. They were able to show that retention of the pulse-labeled RNA on the column depended upon cross-linking, was lost if a photoreversal was done before the material was bound to the column, and did not occur to the same extent if a nonimmune antibody was used.

The secondary structure of a precursor ribosomal RNA (rRNA) in the precursor particle from *Neurospora crassa* mitochondria was investigated by Wollenzien et al.[144] This precursor RNA (35S) contains a 2.2 kbar intervening sequence that is normally spliced out to produce a 25S mature rRNA; however, the precursor accumulates in cells that contain a temperature-sensitive mutation in a nuclear gene.[145] Cross-linking the RNA with aminomethyltrimethylpsoralen in the particle followed by analysis by electron microscopy revealed a large open loop that was formed by cross-linking between sites very close to the left and right splice junctions of the intervening sequence. When the deproteinized RNA was examined under the same conditions, a central hairpin was seen, but not the large intron loop. To demonstrate that these two conformations involved the same sequences at the left splice junction, monoadducts were placed in the RNA in the particle with 390-nm light, and the RNA was deproteinized and then cross-linked under conditions that would favor the intron loop; instead the central hairpin was cross-linked. The experiments indicate that assembly into the precursor particle significantly influences the conformation of the pre-rRNA.

The secondary structure of an RNA molecule containing an intervening sequence was studied during an in vitro splicing reaction.[146] The precursor, a fragment of the mouse insulin I gene containing a 120-nucleotide intron, was made by in vitro transcription. The occurrence of cross-links was detectable by electrophoresis on a urea polyacrylamide gel because cross-linking produced molecules that had reduced mobility. Before the in vitro splicing reaction was started, the pre-RNA had a definite structure, as evidenced by a series of distinct bands on the polyacrylamide gel. The identity of the cross-links was determined by eluting the cross-linked RNAs from the gel and performing reverse transcription with them using synthetic oligonucleotides as primers. Nine of the eleven most frequent cross-links in the molecule were found by this method. They were clustered in the region of the molecule surrounding the intervening sequence. When a nuclear extract from HeLa cells was added to the pre-RNA, very rapidly these specific cross-links were no longer made. This indicates that the extract contains a duplex-destabilizing activity. Psoralen monoadducts were still made on the RNA after incubation in the extract. It was not determined what residual internal structure, if any, the pre-RNA had during splicing.

The base-paired interaction between snRNAs U4 and U6 in intact U4-U6 ribonucleoprotein particles has been studied by psoralen cross-linking by Rinke et al.[147] A mixture of small nuclear ribonucleoproteins U1 to U6 from HeLa cells were purified under nondenaturing conditions on an immunoaffinity column that contains an antibody against the trimethyl-guanosine cap of small nuclear RNAs. These were psoralen cross-linked, deproteinized, and electrophoresed under denaturing conditions. A high-molecular-weight species was present; this proved to be composed of snRNA U4 and U6, as demonstrated by two-dimensional electrophoresis with a photoreversal step before the second dimension. The identity of the cross-linking site was determined by complete digestion of the high-molecular-weight species with ribonuclease T1 and two-dimensional gel electrophoresis that separates the cross-linked oligonucleotides from the non-cross-linked. Two pairs of cross-linked oligonucleotides were identified; these can be aligned to show an interaction between U4 and U6 that contains eight contiguous base pairs. The cross-links between U4 and U6 could be made in the phenolized RNA if the phenolization was done at 0°C, but could not be made if the phenolization was done at 65°C and the RNA was returned to 0°C. This suggests that the contact between the molecules was stable once made but that possibly other secondary structures for the molecules prevented reassociation. The results were used to propose improved models for the secondary structures of U4 and U6 that would accommodate the intermolecular interaction.

IV. CONCLUSION

The use of the psoralens as probes for structure in nucleic acids is providing a wealth of information for a variety of different problems. The ability to obtain useful and detailed information is becoming easier and faster because of improvements in manipulating and sequencing nucleic acids. It can be expected that innovations in the technology will continue to be made. Becker and Wang[148] have described a chemical degradation method that cleaves a DNA molecule at the site of a psoralen adduct. This should allow improved strategies in both DNA and RNA experiments. Cimino et al.[14] have discussed the possibility of directing psoralen monoadducts to specific locations on a target molecule via a complementary synthetic oligonucleotide. A targeting technique like this would greatly simplify the pattern of cross-linking in large molecules and would allow the efficient use of psoralen derivatives that reach long distances.[10,11] These improvements will allow us to learn much more about the higher-order structures of complex nucleic acid molecules.

ACKNOWLEDGMENT

This work was supported by National Institutes of Health grant GM35410.

REFERENCES

1. **Scott, B. R., Pathak, M. A., and Mohn, G. R.,** Molecular and genetic basis of furocoumarin reactions, *Mutat. Res.,* 39, 29, 1976.
2. **Hearst, J. E., Isaacs, S. T., Kanne, D., Rapoport, H., and Straub, K.,** The reaction of psoralens with deoxyribonucleic acids, *Q. Rev. Biophys.,* 17, 1, 1984.
3. **Dall'Acqua, F. S., Marciani, L., Ciavetta, and Rodighiero, G.,** Formation of interstrand cross-linkings in the photoreactions between furocoumarins and DNA, *Z. Naturforsch.,* 26b, 561, 1971.
4. **Wiesehahn, G. and Hearst, J. E.,** DNA unwinding induced by photoaddition of psoralen derivatives and determination of dark-binding equilibrium constants by gel electrophoresis, *Proc. Natl. Acad. Sci. U.S.A.,* 75, 2703, 1978.
5. **Yoakum, G. H. and Cole, R. S.,** Cross-linking and relaxation of supercoiled DNA by psoralen and light, *Biochim. Biophys. Acta,* 521, 529, 1978.
6. **Shen, C.-K.J., Hsieh, T.-S., Wang, J., and Hearst, J. E.,** Photochemical cross-linking of DNA-RNA helices by psoralen derivatives, *J. Mol. Biol.,* 116, 661, 1977.

7. **Thompson, J. F., Bachellerie, J.-P., Hall, K., and Hearst, J. E.,** Dependence of 4'-(hydroxymethyl)-4,5',8-trimethylpsoralen addition on the conformation of ribonucleic acid, *Biochemistry*, 21, 1363, 1982.

8. **Isaacs, S. T., Shen, C.-K. J., Rapoport, H., and Hearst, J. E.,** Synthesis and characterization of new psoralen derivatives on dark binding and photoreactivity, *Biochemistry*, 17, 1251, 1977.

9. **Saffran, W. A., Goldenberg, M., and Cantor, C. R.,** Site-directed psoralen cross-linking of DNA, *Proc. Natl. Acad. Sci. U.S.A.*, 79, 4594, 1982.

10. **Schwartz, D. C., Saffran, W., Welsh, J., Haas, R., Goldenberg, M., and Cantor, C. R.,** New techniques for purifying large DNAs and studying their properties and packaging, *Cold Spring Harbor Symp. Quant. Biol.*, 47, 189, 1983.

11. **Elsner, H., Buchardt, O., Moller, J., and Nielsen, P. E.,** Photochemical crosslinking of protein and DNA in chromatin, synthesis and application of psoralen-cystamine-arylazido photocrosslinking reagents, *Anal. Biochem.*, 149, 575, 1985.

12. **Straub, K., Kanne, D., Hearst, J. E., and Rapoport, H.,** Isolation and characterization of pyrimidine-psoralen photoadducts from DNA, *J. Am. Chem. Soc.*, 103, 2347, 1981.

13. **Kanne, D., Straub, K., Rapoport, H., and Hearst, J. E.,** Psoralen-deoxyribonucleic acid photoreaction, characterization of the monoaddition products from 8-methoxypsoralen and 4,5',8-trimethylpsoralen, *Biochemistry*, 21, 861, 1982.

14. **Cimino, G. D., Gamper, H. B., Isaacs, S. T., and Hearst, J. E.,** Psoralens as photoactive probes of nucleic acid structure and function; organic chemistry, photochemistry, and biochemistry, *Ann. Rev. Biochem.*, 54, 1151, 1985.

15. **Song, P. S. and Tapley, K. J.,** Photochemistry and photobiology of psoralens, *Photochem. Photobiol.*, 29, 1177, 1979.

16. **Parsons, B. J.,** Psoralen photochemistry, *Photochem. Photobiol.*, 32, 813, 1980.

17. **Musajo, L., Rodighiero, G., Colombo, G., Torolone, V., and Dall'Acqua, F. S.,** Photosensitizing furocoumarins: photochemical interaction vs. DNA and −014 CH₃ bergapten (5-methoxypsoralen), *Photochem. Photobiol.*, 5, 739, 1966.

18. **Krauch, C. H., Kramer, D. M., and Wacker, A.,** Zum Wirkungmechanismus Photodyamischer Furocumarine, Photoreaktion von [₄₋₁₄C]Psoralen mit DNS, RNS, Homopolynucleotide and Nucleosiden, *Photochem. Photobiol.*, 6, 341, 1967.

19. **Musajo, L., Bordin, F., and Bevilacqua, R.,** Photoreactions at 3655 A linking the 3-4 double bond of furocoumarins with pyrimidine bases, *Photochem. Photobiol.*, 6, 927, 1967.

20. **Pathak, M. A. and Kramer, D. M.,** Photosensitization of skin in vivo by furocoumarins (psoralens), *Biochim. Biophys. Acta*, 195, 197, 1969.

21. **Musajo, L., Bordin, F., Caporape, G., Marciani, S., and Rigatti, G.,** Photoreactions at 3655 A between pyrimidine bases and skin-photosensitizing furocoumarins, *Photochem. Photobiol.*, 6, 711, 1967.

22. **Dall'Acqua, F. S., Marciani, S., and Rodighiero, G.,** Inter-strand crosslinkages occurring in the photoreactivation between psoralen and DNA, *FEBS Lett.*, 9, 121, 1970.

23. **Cole, R. S.,** Light induced cross linking of DNA in the presence of furocoumarin psoralen, studies with phage lambda, *Escherichia coli*, and mouse leukemic cells, *Biochim. Biophys. Acta*, 217, 30, 1970.

24. **Cole, R. S.,** Inactivation of *Escherichia coli*, F' episomes at transfer, and bacterphage lambda by psoralen plus 360 nm light: significance of deoxyribonucleic acid cross-links, *J. Bacteriol.*, 107, 846, 1971.

25. **Dall'Acqua, F., Marciani, S., and Rodighiero, G.,** Photoreactivity (3655 A) of various skin-photosensitizing furocoumarins with nucleic acids, *Z. Naturforsch.*, 24b, 307, 1969.

26. **Hanson, C. V., Shen, C.-K. J., and Hearst, J. E.,** Cross-linking of DNA in situ as a probe for chromatin structure, *Science*, 193, 62, 1976.

27. **Weisehahn, G. P., Hyde, J. E., and Hearst, J. E.,** The photoaddition of trimethylpsoralen to *Drosophila melanogaster* nuclei: a probe for chromatin substructure, *Biochemistry*, 16, 925, 1977.

28. **Cech, T. and Pardue, M. L.,** Cross-linking of DNA with trimethylpsoralen is a probe for chromatin structure, *Cell*, 11, 631, 1977.

29. **Cech, T., Potter, D., and Pardue, M. L.,** Electron microscopy of DNA cross-linked with trimethylpsoralen: a probe for chromatin structure, *Biochemistry*, 16, 63, 1977.

30. **Hallick, L. M., Yokata, H. A., Bartholomew, J. C., and Hearst, J. E.,** Photochemical addition of the cross-linking reagent 4,5',8-trimethylpsoralen (trioxsalen) to intracellular and viral simian virus 40 DNA-histone complexes, *J. Virol.*, 27, 127, 1978.

31. **Shen, C.-K. J. and Hearst, J. E.,** Chromatin structures of main-band and satellite DNAs in *Drosophila melanogaster* nuclei as probed by photochemical cross-linking of DNA with trioxsalen, *Cold Spring Harbor Symp. Quant. Biol.*, 42, 179, 1977.

32. **Conconi, A., Losa, R., Koller, Th., and Sogo, J. M.,** Psoralen-crosslinking of soluble and of H1-depleted soluble rat liver chromatin, *J. Mol. Biol.*, 178, 920, 1984.

33. **Erard, M., Das, G. C., deMarucia, G., Mazen, A., Pouvet, J., Champagne, M., and Duane, M.,** Ethidium bromide binding to core particle: comparison with native chromatin, *Nucleic Acids Res.*, 6, 3231, 1978.

34. **Labhart, P. and Koller, Th.**, Structural changes of soluble rat liver chromatin induced by the shift in pH from 7 to 9, *Eur. J. Cell. Biol.*, 24, 309, 1981.

35. **Losa, R., Thoma, F., and Koller, Th.**, Involvement of the globular domain of histone H1 in the higher order structures of chromatin, *J. Mol. Biol.*, 175, 529, 1984.

36. **Hyde, J. E. and Hearst, J. E.**, Binding of psoralen derivatives to DNA and chromatin: influence of the ionic environment on dark binding and photoreactivity, *Biochemistry*, 17, 1251, 1977.

37. **Sinden, R. R., Carlson, J. O., and Pettijohn, D. E.**, Torsional tension in the DNA helix measured with trimethylpsoralen in living *E. coli* cells: analogous measurements in insect and human cells, *Cell*, 21, 773, 1980.

38. **Mitra, S., Sen, D., and Crothers, D. M.**, Orientation of nucleosomes and linker DNA in calf thymus chromatin determined by photochemical dichroism, *Nature (London)*, 308, 247, 1984.

39. **Robinson, G. W. and Hallick, L. M.**, Mapping the *in vivo* arrangement of nucleosomes on simian virus 40 chromatin by the photoaddition of radioactive hydroxymethyltrimethylpsoralen, *J. Virol.*, 41, 78, 1983.

40. **Kondoleon, S. K., Robinson, G. W., and Hallick, L. M.**, SV40 virus particles lack a psoralen-accessible origin and contain an altered nucleoprotein structure, *Virology*, 129, 261, 1983.

41. **Gutai, M. and Nathans, D.**, Evolutionary variants of simian virus 40: nucleotide sequence of a conserved SV40 DNA segment containing the origin of viral DNA replication as an inverted repetition, *J. Mol. Biol.*, 126, 259, 1978.

42. **Subramanian, K. and Shenk, T.**, Definition of the boundaries of the origin of DNA replication of simian virus 40, *Nucl. Acids Res.*, 5, 3635, 1978.

43. **Jessel, D., Landau, T., Hudson, J., Lalor, T., Tenen, D., and Livingston, D. M.**, Identification of regions of the SV40 genome which contain preferred SV40 T-antigen binding sites, *Cell*, 8, 535, 1976.

44. **Reed, S., Ferguson, J., Davis, R. W., and Stark, G. R.**, T-antigen binds to simian virus 40 DNA at the origin of replication, *Proc. Natl. Acad. Sci. U.S.A.*, 72, 1605, 1975.

45. **Shortel, D. R., Margolskee, R. F., and Nathans, D.**, Mutational analysis of the simian virus 40 replicon: pseudorevertants of mutants with a defective replication origin, *Proc. Natl. Acad. Sci. U.S.A.*, 76, 6128, 1979.

46. **Tjian, R.**, The binding site on SV40 DNA for a T-antigen-related protein, *Cell*, 13, 165, 1978.

47. **Reddy, V. B., Thimmappaya, B., Dahr, R., Subramanian, K. N., Zain, B. S., Pan, J., Ghosh, P. K., Celma, M. L., and Weissman, S. M.**, The genome of simian virus 40, *Science*, 200, 494, 1978.

48. **Jakobovits, E. B., Bratosin, S., and Aloni, Y.**, A nucleosome-free region in SV40 minichromosomes, *Nature (London)*, 285, 263, 1980.

49. **Saragosti, S., Moyne, G., and Yaniv, M.**, Absence of nucleosomes in a fractoin of SV40 chromatin between the origin of replication and the region coding for the late leader RNA, *Cell*, 20, 65, 1980.

50. **Scott, W. A. and Wigmore, D. J.**, Sites in simian virus 40 chromatin which are preferentially cleaved by endonucleases, *Cell*, 15, 1511, 1978.

51. **Sudin, O. and Varshavsky, A.**, Staphyloccal nuclease makes a single non-random cut in the simian virus 40 minichromosome, *J. Mol. Biol.*, 132, 535, 1979.

52. **Varshavsky, A. J., Sundin, O. H., and Bohn, M. J.**, A stretch of late SV40 viral minichromosome: preferential exposure of the origin of replication as probed by restriction endonucleases, *Nucleic Acids Res.*, 5, 3469, 1978.

53. **Varshavsky, A. J., Sundin, O., and Bohn, M.**, A stretch of late SV40 viral DNA about 400 bp long which includes the origin of replication is specifically exposed in SV40 minichromosomes, *Cell*, 16, 453, 1979.

54. **Waldeck, W., Fohring, B., Chowdhury, K., Gruss, P., and Sauer, G.**, Origin of DNA replication of papovavirus chromatin is recognized by endogenous endonuclease, *Proc. Natl. Acad. Sci. U.S.A.*, 75, 5964, 1978.

55. **Moyne, G., Harper, F., Saragosti, S., and Yaniv, M.**, Absence of nucleosomes in a histone-containing nucleoprotein complex obtained by dissociation of purified SV40 virions, *Cell*, 30, 123, 1982.

56. **Carlson, J. O., Pfenninger, O., Sinden, R. R., Lehman, J. M., and Pettijohn, D. E.**, New procedure using a psoralen derivative for analysis of nucleosome associated DNA sequences in chromatin of living cells, *Nucleic Acids Res.*, 10, 2043, 1982.

57. **Caron, D., Fostl, J., and Rouviere-Yaniv, J.**, Characterization of a histone-like protein extracted from yeast mitochondria, *Proc. Natl. Acad. Sci. U.S.A.*, 76, 4265, 1979.

58. **Kuroiwa, T., Kawano, S., and Hizume, M.**, A method of isolation of mitochondrial nucleod of *Physarum polycephalum* and evidence for the presence of a basic protein, *Exp. Cell Res.*, 97, 435, 1976.

59. **Van Tuyle, G. C. and McPhereson, M. L.**, A compact form of rat liver mitochondrial DNA stabilized by bound proteins, *J. Biol. Chem.*, 254, 6044, 1979.

60. **Potter, D. A., Fostel, J. M., Berninger, M., Pardue, M. L., and Cech, T. R.**, DNA-protein interactions in the *Drosophila melanogaster* mitochondrial genome as deduced from trimethylpsoralen crosslinking patterns, *Proc. Natl. Acad. Sci. U.S.A.*, 77, 4118, 1980.

61. **Pardue, M. L., Fostel, J. M., and Cech, T. R.,** DNA-protein interactions in the *Drosophila virilis* mitochondrial chromosome, *Nucleic Acids Res.,* 12, 1991, 1984.

62. **Fauron, C. M.-R. and Wolstenholme, D. R.,** Structural heterogeniety of mitochondrial DNA molecules within the genus *Drosophila, Proc. Natl. Acad. Sci. U.S.A.,* 73, 3623, 1976.

63. **DeFrancesco, L. and Attardi, G.,** In situ photochemical crosslinking of HeLa cell mitochondrial DNA by a psoralen derivative reveals a protected region near the origin of replication, *Nucl. Acids Res.,* 9, 6017, 1981.

64. **Cech, T. R. and Karrer, K. M.,** Chromatin structure of the ribosomal RNA genes of *Tetrahymena thermophila* as analyzed by trimethylpsoralen crosslinking *in vivo, J. Mol. Biol.,* 136, 395, 1980.

65. **Sogo, J. M., Ness, P. J., Widmer, R. M., Parish, R. W., and Koller, T. H.,** Psoralen-crosslinking of DNA as a probe for the structure of active nucleolar chromatin, *J. Mol. Biol.,* 178, 897, 1984.

66. **Doerig, C., McMaster, G., Sogo, J., Bruggmann, H., and Beard, P.,** Nucleoprotein complexes of minute virus of mice have a distinct structure different from that of chromatin, *J. Virol.,* 58, 817, 1986.

67. **Leinbach, S. S. and Summers, W. C.,** The structure of herpes simplex virus type I DNA as probed by microccocal nuclease digestion, *J. Gen. Virol.,* 51, 45, 1980.

68. **Sinden, R. R., Pettijohn, D. E., and Francke, B.,** Organization of herpes simplex virus type 1 deoxy-ribonucleic acid during replication probed in living cells with 4,5',8-trimethylpsoralen, *Biochemistry,* 21, 4484, 1982.

69. **Cleaver, J. E.,** Chromatin dynamics, fast and slow modes of nucleosome movements revealed through psoralen binding and repair, *Biochim. Biophys. Acta,* 824, 163, 1985.

70. **Mathis, G. and Althaus, F. R.,** Periodic changes of chromatin organization associated with rearrangement of repair patches accompany DNA excision repair of mammalian cells, *J. Biol. Chem.,* 261, 5758, 1986.

71. **Cech, T. R. and Pardue, M. L.,** Electron microscopy of DNA crosslinked with trimethylpsoralen: test of the secondary structure of eukaryotic inverted repeat sequences, *Proc. Natl. Acad. Sci. U.S.A.,* 73, 2644, 1976.

72. **Sinden, R. R., Broyles, S. S., and Pettijohn, D. E.,** Perfect palindromic *lac* operator DNA sequence exists as a stable cruciform structure in supercoiled DNA *in vitro* but not *in vivo, Proc. Natl. Acad. Sci. U.S.A.,* 80, 1797, 1983.

73. **Sinden, R. R. and Pettijohn, D. E.,** Cruciform transitions in DNA, *J. Biol. Chem.,* 259, 6593, 1984.

74. **Hsu, M-T.,** Electron microscopic evidence for the cruciform structure in intracellular SV40 DNA, *Virology,* 143, 617, 1985.

75. **Caffarelli, E., Leoni, L., Sampaolese, B., and Savino, M.,** Persistence of cruciform structure and preferential location of nucleosomes on some regions of pBR322 and ColE1 DNAs, *Eur. J. Biochem.,* 156, 335, 1986.

76. **Shen, C.-K. J. and Hearst, J. E.,** Photochemical cross-linking of transcription complexes with psoralen. I. Covalent attachment of *in vitro* SV40 nascent RNA to its double stranded DNA template, *Nucleic Acids Res.,* 5, 1429, 1978.

77. **Sargan, D. R. and Butterworth, P. H. W.,** Eukaryotic ternary transcription complexes. I. The release of ternary transcription complexes of RNA polymerase I and II by endogenous nucleases of rat liver nuclei, *Nucleic Acids Res.,* 10, 4641, 1982.

78. **Sargan, D. R. and Butterworth, P. H. W.,** Eukaryotic ternary transcription complexes: transcription complexes of RNA polymerase II are with associated histone-containing, nucleosome-like particles *in vivo, Nucleic Acids Res.,* 13, 3805, 1985.

79. **Revet, B. and Benichou, D.,** Electron microscopy of AD5 replicating molecules after *in vivo* photocros-slinking with trioxsalen, *Virology,* 114, 60, 1981.

80. **Richards, O. C., Martin, S. C., Jense, H. G., and Ehrenfeld, E.,** Structure of poliovirus replicative intermediate RNA, electron microscope analysis of RNA cross-linked *in vivo* with psoralen derivative, *J. Mol. Biol.,* 173, 325, 1984.

81. **Meyer, J., Lundquist, R. E., and Maizel, J. V.,** Structural studies of the RNA component of the poliovirus replication complex, characterization by electron microscopy and autoradiography, *Virology,* 85, 434, 1978.

82. **Wittig, B. and Wittig, S.,** Stabilization of R loop structures by photochemical crosslinking with 4.5',8-trimethylpsoralen: application to gene enrichment and molecular cloning, *Biochem. Biophys. Res. Commun.,* 91, 554, 1979.

83. **Chatterjee, P. K. and Cantor, C. R.,** Preparation of psoralen-cross-linked R loops and generation of large deletions by their repair *in vivo, J. Biol. Chem.,* 257, 9173, 1982.

84. **Kaback, D. B., Angerer, L. M., and Davidson, N.,** Improved methods for the formation and stabilization of R-loops, *Nucleic Acids Res.,* 6, 2499, 1979.

85. **Wollenzien, P. L. and Cantor, C. R.,** Marking the polarity of RNA molecules for electron microscopy by covalent attachment of psoralen-DNA restriction fragments, *Proc. Natl. Acad. Sci. U.S.A.,* 79, 3940, 1982.

86. **Kasamatsu, H., Robberson, D. L., and Vinograd, J.,** A novel closed-circular mitochondrial DNA with properties of a replicating intermediate, *Proc. Natl. Acad. Sci. U.S.A.,* 68, 2252, 1971.

87. **Robberson, D. L. and Clayton, D. A.,** Replication of mitochondrial DNA in mouse L cells and their thymidine kinase derivatives: displacement replication on a covalently-closed circular template, *Proc. Natl. Acad. Sci. U.S.A.,* 69, 3810, 1972.

88. **Brown, W. M. and Vinograd, J.,** Restriction endonuclease cleavage maps of animal mitochondrial DNAs, *Proc. Natl. Acad. Sci. U.S.A.,* 71, 4617, 1974.

89. **Peckler, S., Graves, B., Kanne, D., Rapoport, H., Hearst, J. E., and Kim, S.-H.,** Structure of a psoralen-thymine monoadduct formed in photoreaction with DNA, *J. Mol. Biol.,* 162, 157, 1982.

90. **Kim, S.-H., Peckler, S., Graves, B., Kanne, D., Rapoport, H., and Hearst, J. E.,** Sharp kink of DNA at psoralen-cross-link site deduced from crystal structure of psoralen-thymine monoadduct, *Cold Spring Harbor Symp. Quant. Biol.,* 47, 361, 1983.

91. **Land, E. J., Rushton, F. A. P., Beddoes, R. L., Bruce, J. M., Cernik, R. J., Dawson, S. C., and Mills, O. S.,** A [2 + 2] photo-adduct of 8-methoxypsoralen and thymine: X-ray crystal structure; a model for the reaction of psoralens with DNA in the phototherapy of psoriasis, *J. Chem. Soc. Chem. Commun.,* 1, 22, 1982.

92. **Sinden, R. R. and Hagerman, P. J.,** Interstrand psoralen cross-links do not introduce appreciable bends in DNA, *Biochemistry,* 23, 6299, 1984.

93. **Wu, H.-M. and Crothers, D. M.,** The locus of sequence-directed and protein-induced DNA bending, *Nature (London),* 308, 509, 1984.

94. **Hagerman, P. J.,** Evidence for the existence of stable curvature of DNA in solution, *Proc. Natl. Acad. Sci. U.S.A.,* 81, 4632, 1984.

95. **Zolan, M. E., Cortopassi, G. A., Smith, C. A., and Hanawalt, P. C.,** Deficient repair of chemical adducts in α DNA of monkey cells, *Cell,* 28, 613, 1982.

96. **Zolan, M. E., Smith, C. A., and Hanawalt, P. C.,** Formation and repair of furocoumarin adducts in ã deoxyribonucleic acid and bulk deoxyribonucleic acid of monkey cells, *Biochemistry,* 23, 63, 1984.

97. **Courey, A. J., Plon, S. E., and Wang, J. C.,** The use of psoralen-modified DNA to probe the mechanism of enhancer action, *Cell,* 45, 567, 1986.

98. **Haas, R., Murphy, R. F., and Cantor, C. R.,** Testing models of the arrangement of DNA inside bacteriophage lambda by cross-linking the packaged DNA, *J. Mol. Biol.,* 159, 71, 1982.

99. **Welsh, J. and Cantor, C. R.,** Protein-DNA cross-linking, *Trends Biochem.,* 505, 1985.

100. **Saffran, W. A. and Cantor, C. R.,** The complete pattern of mutagenesis arising from the repair of site-specific psoralen crosslinks: analysis by oligonucleotide hybridization, *Nucl. Acids Res.,* 12, 9237, 1984.

101. **Shen, C.-K. J. and Hearst, J. E.,** Psoralen-cross-linked secondary structure map of single-stranded virus DNA, *Proc. Natl. Acad. Sci. U.S.A.,* 73, 2649, 1976.

102. **Shen, C.-K. J., Ikoku, A., and Hearst, J. E.,** A specific DNA orientation in the filamentous bacteriophage fd as probed by psoralen cross-linking and electron microscopy, *J. Mol. Biol.,* 127, 163, 1979.

103. **Ikoku, A. S. and Hearst, J. E.,** Identification of a structural hairpin in the filamentous chimeric phage M13Goril, *J. Mol. Biol.,* 151, 245, 1981.

104. **Peterlin, B. M., Sullivan, M., Westphal, H., and Maizel, J. V., Jr.,** Secondary structure map of psoralen-cross-linked adenovirus DNA studied by electron microscopy, *Virology,* 86, 391, 1978.

105. **Pathak, M. A., Kramer, D. M., and Fitzpatrick, T. B.,** in *Sunlight and Man: Normal and Abnormal Photobiologic Responses,* Fitzpatrick, T. B., Pathek, M. A., Haber, L. C., Seiji, M., and Kukita, A., Eds., University of Tokyo Press, Tokyo, 1974, 335.

106. **Ou, C.-N. and Song, P.-S.,** Photobinding of 8-methoxypsoralen to transfer RNA and 5-fluorouracil-enriched transfer RNA, *Biochemistry,* 17, 1054, 1978.

107. **Bachellerie, J.-P., Thompson, J. F., Wegnez, M. R., and Hearst, J. E.,** Identification of the modified nucleotides produced by covalent photoaddition of hydroxymethyltrimethylpsoralen to RNA, *Nucleic Acids Res.,* 9, 2207, 1981.

108. **Bachellerie, J.-P. and Hearst, J. E.,** Specificity of the photoreaction of 4'-(hydroxymethyl)-4,5',8-trimethylpsoralen with ribonucleic acid, identification of reactive sites in *Escherichia coli* phenylalanine-accepting transfer ribonucleic acid, *Biochemistry,* 21, 1357, 1982.

109. **Thompson, J. F., Wegnez, M. R., and Hearst, J. E.,** Determination of the secondary structure of *Drosophila melanogaster* 5S RNA by hydroxymethyltrimethylpsoralen cross-linking, *J. Mol. Biol.,* 147, 417, 1981.

110. **Youvan, D. C. and Hearst, J. E.,** Sequencing psoralen photochemically reactive sites in *Escherichia coli* 16S rRNA, *Anal. Biochem.,* 119, 86, 1982.

111. **Garrett-Wheeler, E., Lockard, R. E., and Kumar, A.,** Mapping of psoralen cross-linked nucleotides in RNA, *Nucleic Acids Res.,* 12, 3405, 1984.

112. **Turner, S., Thompson, J. F., Hearst, J. E., and Noller, H. F.,** Identification of a site of psoralen cross-linking in *E. coli* 16S ribosomal RNA, *Nucleic Acids Res.,* 10, 2839, 1982.

113. **Nielsen, P. E. and Leick, V.,** Specific photoreactions between psoralens and yeast-tRNA^{Phe}, *Biochem. Biophys. Res. Commun.,* 106, 179, 1982.

114. **Nielsen, P. and Leick, V.**, Photoreaction of 8-methoxypsoralen with yeast-tRNA^{Phe}, identification of the major reaction sites, *Eur. J. Biochem.*, 152, 619, 1985.

115. **Rabin, D. and Crothers, D. M.**, Analysis of RNA secondary structure by photochemical reversal of psoralen cross-links, *Nucl. Acids Res.*, 7, 689, 1979.

116. **Fox, G. E. and Woese, C. R.**, 5S RNA secondary structure, *Nature (London)*, 256, 505, 1975.

117. **Tinoco, I., Jr., Borer, P. N., Dengler, B., Levine, M. D., Uhlenbeck, O. C., Crothers, D. M., and Gralla, J.**, Improved estimation of secondary structures in ribonucleic acids, *Nature (London) New Biol.*, 246, 40, 1973.

118. **Borer, P. N., Dengler, B., Tinoco, I., Jr., and Uhlenbeck, O. C.**, Stability of ribonucleic acid double-stranded helices, *J. Mol. Biol.*, 86, 843, 1974.

119. **Nishikawa, K. and Takemura, S.**, Nucleotide sequence of 5S RNA from *Torulopsis utilis*, *FEBS Lett.*, 40, 106, 1974.

120. **Wollenzien, P. L., Youvan, D. C., and Hearst, J. E.**, Structure of psoralen-cross-linked ribosomal RNA from *Drosophila melanogaster*, *Proc. Natl. Acad. Sci. U.S.A.*, 75, 1642, 1978.

121. **Wollenzien, P. L., Hearst, J. E., Thammana, P., and Cantor, C. R.**, Base-pairing between distant regions of the *Escherichia coli* 16S ribosomal RNA in solution, *J. Mol. Biol.*, 135, 255, 1979.

122. **Thammana, P., Cantor, C. R., Wollenzien, P. L., and Hearst, J. E.**, Cross-linking studies on the organization of the 16S ribosomal RNA within the 30S *Escherichia coli* ribosomal subunit, *J. Mol. Biol.*, 135, 271, 1979.

123. **Wollenzien, P. L., Hearst, J. E., Squires, C., and Squires, C.**, Determining the polarity of the map of cross-linked interactions in *Escherichia coli* 16S ribosomal RNA, *J. Mol. Biol.*, 135, 285, 1979.

124. **Noller, H. F. and Woese, C. R.**, Secondary structure of 16S ribosomal RNA, *Science*, 212, 403, 1981.

125. **Wollenzien, P. L. and Cantor, C. R.**, Gel electrophoresis technique for separating cross-linked RNAs, application to improved electron microscopy analysis of psoralen cross-linked 16S ribosomal RNA, *J. Mol. Biol.*, 159, 151, 1982.

126. **Wollenzien, P. L., Murphy, R. F., Cantor, C. R., Expert-Bezancon, A., and Hayes, D.**, Structure of the *E. coli* 16S rRNA: psoralen and N-acetyl-N'-(p-glyoxylbenzoyl)-cystamine cross-links detected by electron microscopy, *J. Mol. Biol.*, 184, 67, 1985.

127. **Chu, Y. G., Wollenzien, P. L., and Cantor, C. R.**, Use of psoralen monoadducts to compare the structure of 16S rRNA in active and inactive 30S ribosomal subunits, *J. Biomol. Struct. Dyn.*, 1, 647, 1983.

128. **Thompson, J. F. and Hearst, J. E.**, Structure of *E. coli* 16S RNA elucidated by psoralen crosslinking, *Cell*, 32, 1355, 1983.

129. **Thompson, J. F. and Hearst, J. E.**, Structure-function relations in *E. coli* 16S RNA, *Cell*, 33, 19, 1983.

130. **Expert-Bezancon, A. and Wollenzien, P. L.**, Three dimensional arrangement of the *E. coli* 16S rRNA, *J. Mol. Biol.*, 184, 53, 1985.

131. **Hui, C.-F. and Cantor, C. R.**, Mapping the location of psoralen crosslinks on RNA by mung bean nuclease sensitivity of RNA-DNA hybrids, *Proc. Natl. Acad. Sci. U.S.A.*, 82, 1381, 1985.

132. **Wollenzien, P. L.**, Isolation and identification of specific RNA cross-links, *Methods Enzymol.*, in press.

133. **Guerrier-Takada, C., Gardiner, K., Marsh, T., Pace, S., and Altman, S.**, The RNA moiety of ribonuclease is the catalytic subunit of the enzyme, *Cell*, 35, 849, 1983.

134. **Currey, K. M., Peterlin, B. M., and Maizel, J. V., Jr.**, Secondary structure of poliovirus RNA: correlation of computer-predicted with electron microscopically observed structure, *Virology*, 148, 33, 1986.

135. **Jacobson, A. B., Good, L., Simonetti, J., and Zuker, M.**, Some simple computational methods to improve the folding of large RNAs, *Nucleic Acids Res.*, 12, 45, 1984.

136. **Calvet, J. P. and Pederson, T.**, Secondary structure of heterogeneous nuclear RNA: two classes of double-stranded RNA in native ribonucleoprotein, *Proc. Natl. Acad. Sci. U.S.A.*, 74, 3705, 1977.

137. **Calvet, J. P. and Pederson, T.**, Nucleoprotein organization of inverted repeat DNA transcripts in heterogeneous nuclear RNA-ribonucleoprotein particles from HeLa cells, *J. Mol. Biol.*, 122, 361, 1978.

138. **Calvet, J. P. and Pederson, T.**, Heterogeneous nuclear RNA double-stranded regions probed in living HeLa cells by crosslinking with the psoralen derivative aminomethyltrioxsalen, *Proc. Natl. Acad. Sci. U.S.A.*, 76, 755, 1979.

139. **Calvet, J. P. and Pederson, T.**, Base-pairing interactions between small nuclear RNAs and nuclear RNA precursors as revealed by psoralen cross-linking *in vivo*, *Cell*, 26, 363, 1981.

140. **Calvet, J. P., Meyer, L. M., and Pederson, T.**, Small nuclear RNA U2 is base-paired to heterogeneous nuclear RNA, *Science*, 217, 456, 1982.

141. **Raj, N. B. K., Ro-Choi, T. S., and Busch, H.**, Nuclear ribonucleoprotein complexes containing U1 and U2 RNA, *Biochemistry*, 14, 4380, 1975.

142. **Lerner, M. R. and Steitz, J. a.**, Antibodies to small nuclear RNAs complexed with proteins are produced by patients with systemic lupus erythematosus, *Proc. Natl. Acad. Sci. U.S.A.*, 76, 4595, 1979.

143. **Setyono, B. and Pederson, T.**, Ribonucleoprotein organization of eukaryotic RNA, XXX, evidence that U1 small nuclear RNA is a ribonucleoprotein when base-paired with pre-messenger RNA *in vivo*, *J. Mol. Biol.*, 174, 285, 1984.

144. **Wollenzien, P. L., Cantor, C. R., Grant, D., and Lambowitz, A.,** Structure of the unspliced 35S rRNA in *Neurospora crassa* mitochondria detected by psoralen cross-linking, *Cell,* 32, 397, 1983.
145. **Mannella, C. A., Collins, R. A., Green, M. R., and Lambowitz, A. M.,** Defective splicing of mitochondrial rRNA in cytochrome-deficient, nuclear mutants of *Neurospora crassa, Proc. Natl. Acad. Sci. U.S.A.,* 76, 2635, 1979.
146. **Wollenzien, P. L., Goswami, P., Szeberenyi, J., Teave, J., and Goldenberg, C. J.,** The secondary structure of a messenger RNA precursor probed with psoralen is melted in an *in vitro* splicing reaction, *Nucleic Acids Res.,* 15, 9279, 1987.
147. **Rinke, J., Appel, B., Digweed, M., and Luhrmann, R.,** Localization of a base-paired interaction between small nuclear RNAs U4 and U6 in intact U4/U6 ribonucleoprotein particles by psoralen cross-linking, *J. Mol. Biol.,* 185, 721, 1985.
148. **Becker, M. M. and Wang, J. C.,** Use of light for footprinting DNA *in vivo, Nature (London),* 309, 682, 1984.

Chapter 6

GENOTOXIC EFFECTS OF PSORALEN

Wilma A. Saffran

TABLE OF CONTENTS

I. MUTAGENESIS

A. Introduction

The widespread use of psoralen plus UV light (PUVA) therapy for the treatment of skin disorders has focused attention on the possible genotoxic effects of this treatment. Psoralen photoreacts with DNA to form covalent adducts with pyrimidines; many psoralen derivatives, including those in clinical use, are bifunctional reagents and form interstrand cross-links as well as monoadducts (see Chapter 1). DNA-damaging agents are in general mutagenic, recombinogenic, and carcinogenic, and psoralen is no exception. Genotoxic effects have been reported in viruses, prokaryotes, and eukaryotes, and an increased risk of cutaneous squamous-cell carcinoma has been found for patients treated with PUVA for psoriasis.[1] Psoralen mutagenesis has been surveyed in several previous reviews.[2-6]

The initial observation by Altenburg of mutation induction in polar cap cells of *Drosophila melanogaster* eggs by 8-methoxypsoralen (8-MOP) plus UVA irradiation[7] was followed by reports of psoralen genotoxicity in a wide range of organisms. Psoralen mutagenesis was found in *Sarcina lutea*,[8] bacteriophage T4,[9] *Escherichia coli*,[10] and *Aspergillus nidulans*,[11,12] while chromosome aberrations were reported in human lymphocytes.[13]

More recent studies have extended the observations of psoralen mutagenicity to mammalian cells in culture.[14] Burger and Simons detected 8-MOP plus UVA -induced mutations in the HGPRT genes of both Chinese hamster V79 cells[15] and human skin fibroblasts.[16] The number of induced mutations per unit dose was constant for both V79 cells and human fibroblasts, indicating that there is a linear relationship between mutation frequency and the product of 8-MOP concentration and UVA exposure. The mutation frequency per unit dose of the Chinese hamster cells was about twice that of the human fibroblasts, perhaps due to the low excision repair capacity of rodent cells. Similar results in Chinese hamster ovary cells were reported by Schenley and Hsi.[17] Arlett et al.[18] observed mutagenesis of several different loci in mouse lymphoma L5178Y cells and V79 cells.

B. Relative Mutagenicity of Monoadducts and Cross-Links

The simultaneous damage to both strands of the DNA duplex produced by psoralen cross-links poses a greater genetic risk than the more limited involvement of monoadducts on only one strand. In repair-proficient cells the major pathway of monoadduct repair is by excision and resynthesis, a process in which a patch of nucleotides on one strand surrounding the damage site is cut out by nucleases, and the resulting single-strand gap is filled in by DNA polymerase, using the undamaged strand as a template. This error-free repair mechanism will not function on cross-links, since no undamaged template strand is available and additional pathways, including mutagenic repair systems, are required.

It has been proposed that the greater difficulty in repairing cross-links renders them more lethal than monoadducts so that, overall, repaired monoadducts play a greater role in mutagenesis than cross-links.[19] However, cells are capable of repairing a portion of the interstrand cross-links, which therefore constitute potentially mutagenic as well as lethal damage. A number of studies have investigated the relative mutagenicity of monoadducts and cross-links, both to examine the roles of error-free and error-prone pathways in the repair of different DNA lesions and to search for new phototherapeutic agents with lower mutagenic potencies than the bifunctional psoralens in current clinical use.

There have been two approaches to evaluating the genotoxicity of monoadducts and cross-links. In the first approach, samples enriched for either monoadducts or cross-links can be prepared and assayed for mutagenesis. Irradiating bifunctional psoralens with DNA at wavelengths greater than 390 nm will produce relatively pure populations of monoadducts[20] containing cycloadditions predominantly at the 4',5' position. After removal of unbound psoralen, a second irradiation at a shorter wavelength will convert many of these monoadducts to cross-links.

Lambda phage exposed to 8-MOP and the longer wavelength light alone were not mutagenized in *E. coli* host cells.[21,22] However, a second irradiation with shorter-wavelength UV light, producing cross-links, resulted in clear plaque mutations. This mutagenesis was dependent on preirradiation of the host cells with far UV light to induce the error-prone SOS repair system (Weigle mutagenesis). Similarly, after an initial irradiation of *Deinococcus radiodurans* cells with 4,5',8-trimethylpsoralen (TMP) at about 400 nm, the mutation frequency was low, but both mutations and cross-links increased with subsequent exposure to 365-nm light.[23]

An alternative method of preparing monoadduct- or cross-link-enriched DNA is based on the two-hit kinetics of cross-link formation. A brief irradiation with UVA light will produce primarily monoadducts. After washing away the unbound psoralen, a second, more prolonged irradiation will convert monoadducts to cross-links. In yeast,[24] *A. nidulans*,[25] *Chlamydomonas reinhardii*,[26] and Chinese hamster V79 cells[27] treated with 8-MOP or 5-methoxypsoralen (5-MOP), the initial UVA doses produced low levels of mutation, while reirradiation stimulated mutation induction.

The split-dose protocol produced a different result in uvrA⁻ *E. coli*, in which an initial high mutation frequency, produced by the first UVA dose, was decreased by further irradiation.[19,28] In these cells, lacking excision repair, cross-links are lethal rather than mutagenic. In wild-type cells the mutation frequency was stimulated slightly by the second irradiation.[28] These comparisons of monoadducts and cross-links, primarily of 8-MOP, indicated that cross-links are more mutagenic than their monoadduct precursors.

A second approach to the investigation of cross-links and monoadducts is to compare mono- and bifunctional psoralen derivatives. Angelicin, an angular furocoumarin, is an isomer of psoralen but, unlike the linear furocoumarins, cannot form DNA cross-links because of geometric constraints. Alternatively, linear but monofunctional furocoumarins have been synthesized by substituting psoralen at the 3,4 cycloaddition site, chemically blocking interstrand cross-linking. The comparison of the mutagenic potential of various monofunctional furocoumarins has as an additional goal the development of photochemotherapeutic agents with high photoactivity but lower genotoxicity than 8-MOP or TMP.

Venturini et al.[29] compared the mutagenicity of psoralens and angelicins in wild-type and repair-deficient strains of *E. coli*. They found that in wild-type *E. coli*, psoralen and 8-MOP were much more mutagenic than angelicin and 4,5'-dimethylangelicin (4,5'-DMA). In uvrA⁻ bacteria, deficient in excision repair, the psoralens were again more mutagenic than the angelicins, but there was less difference between the two classes of furocoumarins. The mutation frequency was higher for all the compounds in the uvrA⁻ than in the wild-type strain. These experiments indicate that cross-links are more mutagenic than monoadducts, but, in the absence of the error-free excision repair pathway, monoadducts as well are repaired by error-prone processes. Tamaro et al.[30] found similar results when comparing 4,5'-DMA with psoralen and the bifunctional reagent carbomethoxydimethyl psoralen in wild-type and uvr⁻ *E. coli*. The presence of R46 plasmid, which carries an error-prone repair system, increased mutagenesis for all the compounds. The monofunctional derivatives 4-methyl-4',5'dihydropsoralen[31] and 4,4'dimethylangelicin[32] are also less mutagenic than bifunctional psoralens in *E. coli*.

A series of angelicin derivatives tested in uvrB⁻ *Salmonella typhimurium* produced a wide range of mutagenicity which was not well correlated with phototoxicity.[33] For instance, 4,5'-DMA and 4-methylangelicin were highly toxic but had moderate mutagenicity, while 5,5'-dimethylangelicin (5,5'-DMA) and the hydrochloride of 4'-aminomethyl-4,5'-DMA were moderately phototoxic but had the highest mutagenicity. Among the angelicin derivatives tested, the authors suggested 4'-hydroxymethyl-4,5'-DMA, with moderate phototoxicity and low genotoxicity, as a candidate for clinical use.

In bacteriophage lambda, angelicin is less mutagenic than 8-MOP, but the angelicin mutation frequency is increased in uvrA⁻ *E. coli* or by preirradiating bacteria with far UV light to induce the error-prone SOS repair system.[21,22] Weigle reactivation, the enhanced survival of damaged phage in SOS-induced *E. coli*, is lower for 8-MOP than for angelicin in wild-type cells, but is increased for angelicin in uvrA⁻ cells.

In yeast, as in bacteria, the monofunctional furocoumarins are less mutagenic than the bifunctional reagents. Averbeck and co-workers have reported that angelicin,[34] 3-carbethoxypsoralen (3-CP),[35] 5,7-dimethoxycoumarin (5,7-DMC),[36] 4,5'-DMA,[37] and two pyridopsoralens[38] induced nuclear mutations at a lower frequency than psoralen and 8-MOP. As a function of treatment dosage, the order of potency is TMP = 5-MOP > 8-MOP > 4,5'-DMA = 3-CP > angelicin > 5,7-DMC.[39] When plotted as function of survival (i.e., mutations per lethal hit), however, there was less difference between the mono- and bifunctional compounds, and in a forward mutation assay to canavanine resistance the mutations per survivor induced by angelicin and 4,5'-DMA were close to those of 8-MOP.

Averbeck further examined the relationship between adduct formation and genotoxic effects of monofunctional and bifunctional psoralens in diploid yeast.[40] The photobinding capacity of these compounds decreased in the order 7-methylpyrido[3,4-c]psoralen (MPP) > 5-MOP = 3-CP > 8-MOP. There were relatively small differences between 5-MOP, 8-MOP, MPP, and pyridopsoralen in the efficiency of nuclear reversion induction when plotted as a function of dosage. However, the bifunctional compounds 5-MOP and 8-MOP were clearly more mutagenic than the monofunctional psoralens when reversions were examined as a function of DNA adducts, with 8-MOP being most effective.

Monofunctional furocoumarins show relatively greater induction of cytoplasmic petite, or mitochondrial, than of nuclear mutations in yeast.[34-41] The angelicins and monofunctional psoralens are more effective than the cross-linking psoralens in petite induction as a function of survival, and some, such as 3-CPs and pyridopsoralen, induce more petites per dose of radiation.

The relatively greater potency of the monofunctional reagents in inducing mitochondrial than in inducing nuclear mutations is related to the differential repair capacity in the two organelles. There is no excision repair of UV or 8-MOP damage in mitochondrial DNA, in contrast to the extensive excision repair in yeast nuclei. This repair deficiency, combined with the large number of monoadducts per lethal hit in comparison with cross-links, means that mitochondria accumulate heavy burdens of monoadducts under conditions where cell survival is relatively high. The result is loss of mitochondrial DNA and appearance of petite colonies.

Grant et al.[42] compared mutation induction by psoralen and angelicin in wild-type and excision repair-deficient yeast. They found that the reversion frequencies of the two compounds were similar in rad3⁻ yeast, although psoralen was more mutagenic than angelicin in the repair-proficient strain. Thus, in yeast as well as in *E. coli*, most monoadducts are removed by excision repair in wild-type cells and are not substrates for mutagenic processes.

Abel and Schimmer,[43] investigating the alga *C. reinhardii*, found higher mutagenicity for 8-MOP and 8-isoamylenoxypsoralen (8-IOP) than for 5'-methylangelicin (5'-MA).

In Chinese hamster V79 cells, 8-MOP is more effective than 3-CPs in inducing mutations to thioguanine resistance, as a function of both dose and survival.[39,44] At equal doses, 5-MOP induces more mutants than 8-MOP, but at equal numbers of psoralen adducts 8-MOP is more mutagenic, perhaps due to the relatively higher proportion of cross-links produced by 8-MOP than 5-MOP.[45] However, the angelicin derivatives 4,5'-DMA[46] and 5-MA[47] were found to be more mutagenic than 8-MOP. The higher mutagenicity of the angelicins in Chinese hamster cells than in the wild-type bacteria and yeast may be related to the low levels of excision repair found in rodents. Error-prone repair processes may play a greater role in monoadduct repair in these species.

The relative mutagenicity of different psoralen compounds is dependent on the biological system examined. For instance, Tamaro et al.[48] found differences between *E. coli* and V79 cells when studying a series of water-soluble 8-methoxy and 5-methoxy psoralen derivatives. While the parent compounds 5-MOP and 8-MOP were most genotoxic in *E. coli*, the 8-methoxy derivatives were most mutagenic in the V79 cells.

Overall, the monofunctional furocoumarins are less mutagenic than the bifunctional psoralens. There is, however, considerable variation among the various derivatives and, especially under conditions where excision repair is deficient or lacking, monoadducts may induce as many mutations as cross-links.

C. Nature of Psoralen Mutations

DNA-damaging agents induce characteristic changes in the sequence of DNA, depending on the sites of reaction, the nature of the lesions, and the mechanism of repair. Possible mutations include base-pair substitutions and additions or deletions of base pairs. The substitutions may be (1) transitions, the change from one pyrimidine to another pyrimidine or one purine to another; or (2) transversions, the interconversion of purines and pyrimidines.

In an early investigation, Drake and McGuire[9] produced mutations in the *rII* locus of bacteriophage T4 and examined the reversion properties of the mutants. A majority of the psoralen-induced mutants were revertible by base analogues and were classified as transitions, while a substantial minority were not revertible and were classified as transversions. There were, in addition, proflavine-revertible frameshift mutations, but, after correction for the substantial frequency of spontaneous frameshifts expected at the *rII* locus, the authors concluded that few or none of these was induced by psoralen.

Igali et al.[10] reported that 8-MOP plus near UV was active in producing reversion and suppressor mutations of *E. coli trp*, both single-base transitions. Kirkland et al.[49] confirmed that both 8-MOP and TMP produce base substitutions in *E. coli* and further reported that they were inactive in the induction of frameshifts in *S. typhimurium*. In contrast, 8-MOP,[50,51] TMP, and two pyridopsoralens induced frameshift mutations in both *E. coli* and *S. typhimurium* in the dark, at high concentrations, while 5-MOP and 3-CPs were inactive in the dark.[52]

Several recent studies have examined the nature of psoralen-induced mutations by direct DNA sequencing of the affected genes. In these investigations *E. coli* host cells were transformed with plasmid or phage DNA molecules which had been photoreacted with psoralen in vitro. After in vivo repair and mutagenesis, mutant DNA was isolated and sequenced.

Saffran and Cantor[53,54] added psoralen site specifically to a 50-base pair (-bp) region of the *tet* gene of plasmid pBR322 by first incorporating mercurated nucleotides at the target region, then directing a sulfhydryl derivative of 4'-aminomethyl-4,5',8-trimethylpsoralen (AMT) to the target through Hg-S linkages. After photoreaction, cross-linked plasmid was isolated and transformed into *E. coli* cells which had been previously irradiated with far UV light to induce the mutagenic SOS DNA repair system. Mutants within the target region were detected by colony hybridization to oligonucleotides; plasmids with sequence changes form imperfect hybrids and are detected by loss of signal at elevated temperatures.

The mutants identified in this manner were primarily transitions, with a minority of transversions. Both transitions and transversions occurred at some sites. The mutations were clustered at a few sites, where T-T or, to a lesser extent, T-C cross-links were likely to form, indicating that mutagenesis is targeted to the sites of psoralen reaction.

Piette et al.[55] studied psoralen mutagenesis in the bacteriophage M13. M13mp10 replicative-form DNA was reacted randomly in vitro with HMT and transformed into *E. coli* cells. The survivors were screened for mutations in the lac gene, and the beginning of the lac regions was sequenced. Most of the changes found were single-base substitutions, but the majority of mutations were transversions rather than transitions. These changes were clustered at a few sites within the promoter region. All occurred at AT sites where T-T or T-C cross-links could have been formed. Both transitions and transversions were found at the same sites. There were, in addition, several single-base frameshifts, both additions and deletions, at runs of Ts or Cs, which may have been of spontaneous origin. Miller and Eisenstadt have also found that angelicin induces primarily transversion in the *lacI* gene of *E. coli*.[102]

Yoon[56] prepared plasmid pBR322 containing TMP within a 30-bp region in the tet promoter by photoreacting a purified restriction fragment in vitro, then ligating the treated fragment with undamaged DNA to reconstitute the plasmid. In contrast to the other investigations, the two mutants with changes in the target region contained incomplete repetitions of the 31-bp restriction fragment, rather than single-base substitutions. This was probably the consequence of imperfect ligation of the modified ends of the DNA, similar to the deletions at ligation sites seen in similar experiments with far UV-damaged plasmids[57] rather than psoralen-induced mutation.

These studies indicate that psoralen plus near UV induces single-base substitutions, both transitions and transversions, and that the mutations are targeted to psoralen reaction sites. There were differences in the relative proportions of transitions and transversions found, which may be due to differences in the psoralen derivatives, the genes analyzed, the presence of SOS induction, or the biological system. Dark mutagenesis by psoralen occurs by a different mechanism, producing frameshifts rather than base substitutions.

D. Dark Mutagenesis of Psoralen

Doses of psoralen producing high mutation frequencies after photodynamic activation have, in general, no mutagenic activity in the absence of UVA light. However, in the treatment of psoriasis, 8-MOP is commonly administered orally, thus exposing patients internally to repeated doses of psoralen in the dark. Several investigators have examined the mutagenic capacity of psoralens in the dark.

8-MOP was reported to induce frameshift mutations at high concentration in *E. coli*[50,51,58] and *S. typhimurium*[51,52] under growth conditions. However, mutagenicity was weak and was not observed in other studies using lower concentrations of 8-MOP or in the absence of DNA replication.[49,59] This dark frameshift mutagenesis, unlike the UVA-dependent base substitutions, was unaffected by uvrA⁻ or uvrB⁻ excision repair deficiencies in *E. coli* and thus may be due to the noncovalent dark complexing of psoralen with DNA.

Similar weak frameshift mutagenicity at high concentrations was observed for angelicin,[58] 5-MOP,[58,60] 4,5′,8-TMP,[52] and two pyridopsoralens,[52] while other studies found no mutagenic activity of TMP,[58] 5-MOP,[49,58] and 3-CPs.[58] Lecointe, assaying the dark ability of psoralens to induce SOS repair functions in *E. coli*, found weak activities for psoralen, TMP, and 8-MOP[61] at high drug concentrations.

Further investigations into the effect of microsomal activation on dark mutagenicity produced conflicting results. Kirkland et al.[49] reported that 8-MOP and TMP produced base substitutions in uvrA⁻ *E. coli* after S9 mix microsomal activation and that this mutagenesis was light-independent. No effect on frameshift activity in *S. typhimurium* was seen. In contrast, Lecointe found that S9 mix reduced SOS induction by TMP but had no effect on psoralen of 8-MOP.[61] Schimmer and Fischer reported a decrease in the number of 8-MOP plus UVA-induced revertants of *C. reinhardii* after treatment with S9 mix.[62] Ivie et al.[63] reported weak mutagenic activities of several psoralen epoxides in the dark.

As all the reported incidence of dark mutagenesis occurred at concentrations far above those observed in plasma during PUVA treatment, it would seem that there is low genetic risk due to dark mutagenesis associated with photochemotherapy.

II. RECOMBINOGENESIS

A. Genetic Recombination

Recombinational repair plays a major role in the cellular response to psoralen photoreaction, particularly interstrand cross-links. In the initial step, the damaged DNA is incised by the enzymes of the excision repair system. The resulting gapped or cut DNA is then repaired by recombinational exchange with undamaged homologous DNA.

In *E. coli*, the uvrABC excision nuclease nicks one strand of the DNA on both sides of the psoralen modification.[64,65] Strand rejoining is accompanied by exchange between homologous duplexes[66] and is reduced or absent in recombination-deficient strains.[67] Strain survival is also correlated with strand rejoining. Sinden and Cole[67] reported that the strains with the highest sensitivity to TMP cross-linking were recA, lexA, recB recC sbcB recF, and recB recL, which were completely deficient in strand rejoining.

Howard-Flanders and co-workers[68-71] have studied psoralen-induced recombination in lambda phage. They found that recombination between infecting lambda phage and resident prophage in *E. coli* lysogens was stimulated by psoralen modification of the infecting phage particles.[68] This exchange followed second-order kinetics, suggesting that cross-links, rather than monoadducts, were responsible. Cassuto et al.[69] compared the recombinogenic effectiveness in this system of TMP, which is an efficient cross-linker, and the furanochrome khellin, which produces monoadducts but few cross-links. They reported that the TMP produced more recombination than khellin at equal levels of total adducts and estimated that cross-links were about 20 times more effective than monoadducts in the induction of genetic exchanges.

Ross and Howard-Flanders[70,71] studied the initial steps in recombination by examining the effect of psoralen-damaged lambda on undamaged phage. When *E. coli* host cells infected with intact, unmodified lambda were superinfected with psoralen-modified lambda, the undamaged DNA was cut. This cutting reaction was dependent of the uvrA and recA genes, and required the presence of psoralen-damaged homologous DNA.[71] Superinfection with nonhomologous phages produced no incision.

Cupido and Bridges[72] also reported that repair of 8-MOP cross-links in *E. coli* and lambda phage was associated with homologous recombination in uvr⁻ strains. They found no cross-link repair in *E. coli* cells containing only a single genome or in lambda at a multiplicity of infection of less than one.

Mammalian viruses also undergo psoralen-induced recombination. Genetic recombination between herpes simplex type I viruses is induced by psoralen treatment.[73] In addition, both herpes simplex[73] and simian virus 40[74] (SV40) DNA photoreacted with TMP showed multiplicity reactivation, the increase in survival associated with multiplicities of infection greater than one. A single cross-link per SV40 particle was lethal at a multiplicity of infection of one.[74]

In yeast, psoralen-modified DNA is incised by the RAD3 excision repair system, producing single-strand breaks after angelicin damage and double-strand breaks after TMP or 8-MOP cross-linking.[75] Strand rejoining is controlled by PSO2[76] and the genes of the RAD52[75] recombination repair system. Psoralen sensitivity is dependent on the ploidy of yeast cells. Diploid yeast are more resistant than haploid cells, and haploid cells are more resistant in S and G2 phases, after DNA replication, than in G1 phase, indicating that homologous recombination is involved in psoralen damage repair.[41]

Psoralen photoreaction induces mitotic recombination, both gene conversion (intragenic) and crossing over (intergenic) in diploid yeast. The relative efficiencies of monoadducts and cross-links in recombinogenesis have been studied by comparing mono- and bifunctional furocoumarins, as well as by split-dose experiments with cross-linking psoralens.

Although psoralen and 8-MOP were found to induce more gene conversion than angelicin and 3-CPs, as a function of dose,[35] the monofunctional pyridopsoralens were as efficient as

5-MOP and 8-MOP.[40] In addition, induced conversions per viable cell were similar for all compounds.[35,40] However, the bifunctional reagents induced higher levels of mitotic crossing over than the monofunctional compounds. Averbeck[40] compared the levels of photobinding and recombinogenesis for 8-MOP, 5-MOP, MPP, and 3-CPs. He reported that recombinants per psoralen adduct were higher for the cross-linkers than the monofunctional psoralens, although the differences were not as great for recombinogenesis as for mutagenesis.

Gene conversion and crossing over were reduced in strains with pso2 mutations for 8-MOP, AMT, and psoralen, but not for 3-CPs, suggesting that the PSO2 gene plays a specific role in cross-link recombinogenesis.[77] Recombination was absent in rad52⁻ mutants for both mono- and bifunctional psoralens.

8-MOP and 4,5'-DMA were also shown to produce mitotic nondisjunction in diploid yeast.[39] Again, 8-MOP induced nondisjunction at lower doses, but the induction per surviving cell was similar for the two compounds.

In split-dose experiments with 8-MOP,[24] the initial irradiation, producing primarily monoadducts, generated low levels of gene conversion. Subsequent reirradiation to produce cross-links from the monoadduct precursors induced high gene conversion frequencies.

Psoralen cross-links thus appear to be highly recombinogenic in yeast, although monoadducts also induce recombination, particularly at high doses. As with mutagenesis, there is considerable variation among the different mono- and bifunctional furocoumarins.

B. Sister Chromatid Exchanges and Chromosome Aberrations

Psoralen induction of sister chromatid exchanges (SCE) has been observed after in vitro treatment of human lymphocytes[78,83] and epidermal cells,[84] as well as Chinese hamster ovary cells.[60,85-86] Cassel and Latt,[86] measuring [³H]8-MOP incorporation and SCE induction in Chinese hamster ovary cells, calculated that 8-MOP induces one SCE per 200 adducts. 5-MOP, TMP, and several 8-MOP and TMP derivatives also induce SCEs in human and hamster cells.[82,83] TMP and its derivatives were more active than the 8-MOP derivatives in SCE induction.

In vivo treatment of Chinese hamsters with 8-MOP plus UVA induced SCEs in cheek pouch mucosal cells.[87] When lymphocytes were taken from psoriasis patients after 8-MOP ingestion and then UVA-irradiated in vitro, an increase in SCE was observed.[88,89] However, several studies failed to detect an increase in SCEs after in vivo PUVA treatment.[88-91] More recently, Bredberg et al.[88] have reported a small increase in SCE levels in lymphocytes, but not skin fibroblasts, taken from patients after 1 to 6 years of PUVA therapy. Skin fibroblasts taken from patients after 5 years of PUVA therapy showed elevated levels of DNA cross-linking. The effective short-term light exposure of lymphocytes during PUVA treatment is apparently not high enough to induce detectable genotoxicity, but damage may accumulate during long-term therapy.   The relative efficiencies of 8-MOP monoadducts and cross-links have been evaluated by reirradiation experiments. In Chinese hamster ovary cells,[92] human fibroblasts,[93] and mouse lymphoma cells,[94] conversion of monoadducts to cross-links stimulated SCEs, indicating that cross-links are more effective than monoadducts. When mouse fibroblasts were incubated between the two light doses to allow monoadduct repair, SCEs decreased to control levels.

SCE induction by mono- and bifunctional furocoumarins has also been examined. 8-MOP and 5-MOP were found to be more efficient than angelicin and 3-CPs.[95-99] However, 5-MA,[97] 5'-MA,[43] and the monofunctional psoralens MPP and PP[98] had high activity, while 8-isoamylenoxypsoralen was inactive.[43] The dose response was different for the monofunctional compounds, which reached plateaus of SCE induction at high doses, than for 8-MOP and 5-MOP, which continued to increase.[97,98]

Psoralen photoreaction has been shown to induce chromosome aberrations in vitro in human lymphocytes,[79,95] Chinese hamster ovary cells,[100] and mouse lymphoma cells.[94,101]

Psoralen plus UVA induces chromosome constrictions, gaps, breaks, and exchanges, as well as micronuclei formed from acentric chromosome fragments. Psoralen, 8-MOP, and 5-MOP were reported to induce aberrations at lower doses than angelicin.[95,100] Reirradiation experiments showed that 8-MOP cross-links are more active than monoadducts in producing micronuclei[94] and a wide variety of chromosomal aberrations, including dicentrics in first-division cells and chromatid deletions and exchanges in second-division cells.[101]

In mixed in vivo-in vitro experiments, lymphocytes which were taken from patients after 8-MOP ingestion and then UVA irradiated were found to have increased levels of chromosomal aberrations.[88,89] However, no elevation in aberrations was detected in lymphocytes from psoriasis patients treated in vivo with PUVA.[88,89,91]

REFERENCES

1. **Stern, R. S., Laird, N., Melski, J., Parrish, J. A., Fitzpatrick, T. B., and Bleich, H. L.**, Cutaneous squamous-cell carcinoma in patients treated with PUVA, *N. Engl. J. Med.*, 310, 1156, 1984.
2. **Scott, B. R., Pathak, M. A., and Mohn, G. R.**, Molecular and genetic basis of furocoumarin reactions, *Mutat. Res.*, 39, 29, 1976.
3. **Song, P. S. and Tapley, K. J., Jr.**, Photochemistry and photobiology of psoralens, *Photochem. Photobiol.*, 29, 1177, 1979.
4. **Grekin, D. A. and Epstein, J. H.**, Psoralens, UVA (PUVA) and photocarcinogenesis, *Photochem. Photobiol.*, 33, 957, 1981.
5. **Rodighiero, G., Dall'Acqua, F., and Pathak, M. A.**, Photobiological properties of monofunctional furocoumarin derivatives, in *Topics in Photomedicine*, Smith, K., Eds., Plenum Press, New York, 1984, 319.
6. **Roelandts, R.**, Mutagenicity and carcinogenicity of methoxsalen plus UV-A, *Arch. Dermatol.*, 120, 662, 1984.
7. **Altenburg, E.**, Studies of the enhancement of mutation rate by carcinogens, *Tex. Rep. Biol. Med.*, 14, 481, 1956.
8. **Mathews, M. M.**, Comparative study of the lethal photosensitization of *Sarcina lutea* by 8-methoxypsoralen and toluidene blue, *J. Bacteriol.*, 85, 322, 1963.
9. **Drake, J. W. and McGuire, J.**, Properties of r mutants of bacteriophage T4 photodynamically induced in the presence of thipyronin and psoralen, *J. Virol.*, 1, 260, 1967.
10. **Igali, S., Bridges, B. A., Ashwood-Smith, M. J., and Scott, B. R.**, Mutagenesis in *E. coli*. IV. Photosensitization to near ultraviolet light by 8-methoxypsoralen, *Mutat. Res.*, 9, 21, 1970.
11. **Alderson, T. and Scott, B. R.**, The photosensitizing effect of 8-methoxypsoralen on the inactivation and mutation of *Aspergillus conidia* by near ultra-violet light, *Mutat. Res.*, 9, 569, 1970.
12. **Scott, B. R. and Alderson, T.**, The random (non-specific) forward mutation of gene loci in *Aspergillus nidulans conidia* after photosensitization to near-ultraviolet light (365 nm) by 8-methoxypsoralen, *Mutat. Res.*, 12, 29, 1971.
13. **Sasaki, M. S. and Tonomura, A.**, A high susceptibility of Fanconi's anemia to chromosome breakage by DNA cross-linking agents, *Cancer Res.*, 33, 1829, 1973.
14. **Arlett, C. F.**, Mutagenesis in cultured mammalian cells, *Stud. Biophys.*, 36/37, 139, 1973.
15. **Burger, P. M. and Simons, J. W.**, Mutagenicity of 8-methoxypsoralen and long-wave ultraviolet irradiation in V79 Chinese hamster cells. A first approach to a risk estimate in photochemotherapy, *Mutat. Res.*, 60, 381, 1979.
16. **Burger, P. M. and Simons, J. W.**, Mutagenicity of 8-methoxypsoralen and long-wave ultraviolet irradiation in diploid human skin fibroblasts: an improved risk estimate in photochemotherapy, *Mutat. Res.*, 63, 371, 1979.
17. **Schenley, R. L. and Hsie, A. W.**, Interaction of 8-methoxypsoralen and near-UV light causes mutation and cytotoxicity in mammalian cells, *Photochem. Photobiol.*, 33, 179, 1981.
18. **Arlett, C. F., Heddle, J. A., Broughton, B. C., and Rogers, A. M.**, Cell killing and mutagenesis by 8-methoxypsoralen in mammalian (rodent) cells, *Clin. Exp. Dermatol.*, 5, 147, 1980.
19. **Seki, T., Nozu, K., and Kondo, S.**, Differential causes of mutation and killing in *Escherichia coli* after psoralen plus light treatment: monoadducts and crosslinks, *Photochem. Photobiol.*, 27, 18, 1978.
20. **Chatterjee, P. K. and Cantor, C. R.**, Photochemical production of psoralen DNA monoadducts capable of subsequent photocross-linking, *Nucleic Acids Res.*, 5, 3619, 1978.

21. **Belogurov, A. A. and Zavilgelsky, G. B.,** Mutagenic effect of furocoumarin monoadducts and crosslinks on bacteriophage lambda, *Mutat. Res.,* 84, 11, 1981.

22. **Zavilgelsky, G. B., Belogurov, A. A., and Kriuger, D. N.,** W-reactivation and W-mutagenesis in lambda and T7 bacteriophages: a comparative study of the action of ultraviolet radiation (254 nm) and of the photosensitizing agents, 8-methoxypsoralen and angelicin, *Genetika,* 18, 24, 1982.

23. **Yatagai, F. and Kitayama, S.,** Mutation induction by crosslinks in DNA of *Deinococcus radiodurans, Biochem. Biophys. Res. Commun.,* 112, 458, 1983.

24. **Cassier, C., Chanet, R., and Moustacchi, E.,** Mutagenic and recombinogenic effects of DNA crosslinks induced in yeast by 8-methoxypsoralen photoaddition, *Photochem. Photobiol.,* 39, 799, 1984.

25. **Scott, B. R. and Maley, M. A.,** Mutagenicity of monoadducts and crosslinks induced in *Aspergillus nidulans* by 8-methoxypsoralen plus 365 nm radiation, *Photochem. Photobiol.,* 34, 63, 1981.

26. **Schimmer, O.,** Effect of reirradiation with UVA on inactivation and mutation induction in arg⁻ cells of *Chlamydomonas reinhardii* pretreated with furocoumarins plus UVA, *Mutat. Res.,* 109, 195, 1983.

27. **Babudri, N., Pani, B., Venturini, S., Tamaro, M., Monti-Bragadin, C., and Bordin, F.,** Mutation induction and killing of V79 Chinese hamster cells by 8-methoxypsoralen plus near ultraviolet light: relative effects of monoadducts and crosslinks, *Mutat. Res.,* 91, 391, 1981.

28. **Bridges, B. A., Mottershead, R. P., and Knowles, A.,** Mutation induction and killing of *Escherichia coli* by DNA adducts and crosslinks: a photobiological study with 8-methoxypsoralen, *Chem. Biol. Interact.,* 27, 221, 1979.

29. **Venturini, S., Tamaro, M., Monti-Bragadin, C., Bordin, F., Baccicchetti, F., and Carlassare, F.,** Comparative mutagenicity of linear and angular furocoumarins in *Escherichia coli* strains deficient in known repair functions, *Chem. Biol. Interact.,* 30, 203, 1980.

30. **Tamaro, M., Monti-Bragadin, C., Rodighiero, P., Baccichetti, F., Carlassare, F., and Bordin, F.,** Killing and mutations induced by mono- and bifunctional furocoumarins in wild type and excision-less *Escherichia coli* strains carrying the R46 plasmid, *Photobiochem. Photobiophys.,* 10, 261, 1986.

31. **Fujita, H.,** Photobiological activity of 4-methylpsoralen and 4-methyl-4',5'-dihydropsoralen with respect to lethal and mutagenic effects on *E. coli* and prophage induction, *Photochem. Photobiol.,* 39, 835, 1984.

32. **Baccichetti, F., Bordin, F., Carlassare, F., Peron, M., Guiotto, A., Rodighiero, P., Dall'Acqua, F., and Tamaro, M.,** 4,4'-Dimethylangelicin, a monofunctional furocoumarin showing high photosensitizing activity, *Photochem. Photobiol.,* 34, 649, 1981.

33. **Venturini, S., Tamaro, M., Monti-Bragadin, C., and Carlassare, R.,** Mutagenicity in *Salmonella typhimurium* of some angelicin derivatives proposed as new monofunctional agents for the photochemotherapy of psoriasis, *Mutat. Res.,* 88, 17, 1981.

34. **Averbeck, D., Chandra, P., and Biswas, R. K.,** Structural specificity in the lethal and mutagenic activity of furocoumarins in yeast cells, *Radiat. Environ. Biophys.,* 12, 241, 1975.

35. **Averbeck, D. and Moustacchi, E.,** Genetic effect of 3-carbethoxypsoralen, angelicin, psoralen and 8-methoxypsoralen plus 365 nm irradiation in *Saccharomyces cerevisiae:* induction of reversions, mitotic crossing-over, gene conversion and cytoplasmic "petite" mutations, *Mutat. Res.,* 68, 133, 1979.

36. **Averbeck, D. and Moustacchi, E.,** Decreased photoinduced mutagenicity of monofunctional as opposed to bifunctional furocoumarins in yeast, *Photochem. Photobiol.,* 31, 475, 1980.

37. **Averbeck, D., Averbeck, S., and Dall'Acqua, F.,** Mutagenic activity of three monofunctional and three bifunctional furocoumarins in yeast *(Saccharomyces cerevisiae), Farmaco,* 36, 492, 1981.

38. **Averbeck, D., Averbeck, S., Bisagni, E., and Moron, L.,** Lethal and mutagenic effects photoinduced in haploid yeast *(Saccharomyces cerevisiae)* by two new monofunctional pyridopsoralens compared to 3-carbethoxypsoralen and 8-methoxypsoralen, *Mutat. Res.,* 148, 47, 1985.

39. **Averbeck, D., Papadopoulo, D., and Quinto, I.,** Mutagenic effects of psoralens in yeast and V79 Chinese hamster cells, *Natl. Cancer Inst. Monogr.,* 66, 127, 1984.

40. **Averbeck, D.,** Relationship between lesions photoinduced by mono- and bi-functional furocoumarins in DNA and genotoxic effects in diploid yeast, *Mutat. Res.,* 151, 217, 1985.

41. **Henriques, J. A. P., Chanet, R., Averbeck, D., and Moustacchi, E.,** Lethality and "petite" mutation induced by the photoaddition of 8-methoxypsoralen in yeast, *Mol. Gen. Genet.,* 158, 63, 1977.

42. **Grant, E. L., von Borstel, R. C., and Ashwood-Smith, M. J.,** Mutagenicity of cross-links and monoadducts of furocoumarins (psoralen and angelicin) induced by 360 nm radiation in excision-repair-defective and radiation-insensitive strains of *Saccharomyces cerevisiae, Environ. Mutagen.,* 1, 55, 1979.

43. **Abel, G. and Schimmer, O.,** Mutagenicity and toxicity of furocoumarins in 2 test systems, *Mutat. Res.,* 90, 451, 1981.

44. **Papadopoulo, D., Sagliocco, F., and Averbeck, D.,** Mutagenic effects of 3-carbethoxypsoralen and 8-methoxypsoralen plus 365 nm irradiation in mammalian cells, *Mutat. Res.,* 124, 287, 1983.

45. **Papadopoulo, D. and Averbeck, D.,** Genotoxic effects and DNA photoadducts induced in Chinese hamster V79 cells by 5-methoxypsoralen and 8-methoxypsoralen, *Mutat. Res.,* 151, 281, 1985.

46. **Swart, R. N., Beckers, M. A., and Schothorst, A. A.,** Phototoxicity and mutagenicity of 4,5'-dimethylangelicin and long-wave ultraviolet irradiation in Chinese hamster cells and human skin fibroblasts, *Mutat. Res.,* 124, 271, 1983.

47. **Loveday, K. S. and Donahue, B. A.,** Induction of sister chromatid exchanges and gene mutations in Chinese hamster ovary cells by psoralens, *Natl. Cancer Inst. Monogr.,* 66, 149, 1984.

48. **Tamaro, M., Bastaldi, S., Carlassare, F., Babudri, N., and Pani, B.,** Genotoxic activity of some water-soluble derivatives of 5-methoxypsoralen and 8-methoxypsoralen, *Carcinogenesis,* 7, 605, 1986.

49. **Kirkland, D. J., Creed, K. L., and Mannisto, P.,** Comparative bacterial mutagenicity studies with 8-methoxypsoralen and 4,5',8-trimethylpsoralen in the presence of near-ultraviolet light and in the dark, *Mutat. Res.,* 116, 73, 1983.

50. **Clarke, C. H. and Wade, M. T.,** Evidence that caffeine, 8-methoxypsoralen and steroidal diamines are frameshift mutagens for *E. coli* K-12, *Mutat. Res.,* 28, 123, 1975.

51. **Bridges, B. A. and Mottershead, R. P.,** Frameshift mutagenesis in bacteria by 8-methoxypsoralen (methoxalen) in the dark, *Mutat. Res.,* 44, 305, 1977.

52. **Quinto, I., Averbeck, D., Moustacchi, E., Hrisoho, Z., and Moron, J.,** Frameshift mutagenicity in *Salmonella typhimurium* of furocoumarins in the dark, *Mutat. Res.,* 136, 49, 1984.

53. **Saffran, W. A. and Cantor, C. R.,** Mutagenic SOS repair of site-specific psoralen damage in plasmid pBR322, *J. Mol. Biol.,* 178, 595, 1984.

54. **Saffran, W. A. and Cantor, C. R.,** The complete pattern of mutagenesis arising from the repair of site-specific psoralen crosslink: analysis by oligonucleotide hybridization, *Nucleic Acids Res.,* 12, 9237, 1984.

55. **Piette, J., Decuyper-Debergh, D., and Gamper, H.,** Mutagenesis of the lac promoter region in M13mp10 phage DNA by 4'-hydroxymethyl-4,5',8-trimethylpsoralen, *Proc. Natl. Acad. Sci. U.S.A.,* 82, 7355, 1985.

56. **Yoon, K.,** Localized mutagenesis of the tetracycline promoter region in pBR322 by 4,5',8-trimethylpsoralen, *Mutat. Res.,* 93, 253, 1982.

57. **Livneh, Z.,** Directed mutagenesis method for analysis of mutagen specificity. Application to ultraviolet-induced mutagenesis, *Proc. Natl. Acad. Sci. U.S.A.,* 80, 237, 1983.

58. **Ashwood-Smith, M. J.,** Frameshift mutations in bacteria produced in the dark by several furocoumarins: absence of activity of 4,5',8-trimethylpsoralen, *Mutat. Res.,* 58, 23, 1978.

59. **Scott, B. R.,** Failure to detect a mutagenic activity of 8-methoxypsoralen (in the dark) in strains of *Salmonella typhimurium* and *Escherichia coli, Mutat. Res.,* 40, 167, 1976.

60. **Ashwood-Smith, M. J., Poulton, G. A., Barker, M., and Mildenberger, M.,** 5-Methoxypsoralen, an ingredient in several suntan preparations, has lethal, mutagenic and clastogenic properties, *Nature (London),* 285, 4007, 1980.

61. **Lecointe, P.,** Induction of the SFIA SOS repair function by psoralens in the dark, *Mutat. Res.,* 131, 111, 1984.

62. **Schimmer, O. and Fischer, K.,** Metabolic inactivation of 8-methoxypsoralen (8-MOP) by rat liver microsomal preparations, *Mutat. Res.,* 79, 327, 1980.

63. **Ivie, G. W., Macgregor, J. T., and Hammock, B. D.,** Mutagenicity of psoralen epoxides, *Mutat. Res.,* 79, 73, 1980.

64. **Sancar, A. and Rupp, W. D.,** A novel repair enzyme: UVR ABC excision nuclease of *Escherichia coli* cuts a DNA strand on both sides of the damaged region, *Cell,* 33, 249, 1983.

65. **Sancar, A., Franklin, K. A., and Sancar, G.,** Repair of psoralen and acetylaminofluorene DNA adducts by ABC excinuclease, *J. Mol. Biol.,* 184, 725, 1985.

66. **Cole, R. S.,** Repari of DNA containing interstrand crosslinks in *Escherichia coli:* sequential excision and recombination, *Proc. Natl. Acad. Sci. U.S.A.,* 70, 1064, 1973.

67. **Sinden, R. R. and Cole, R. S.,** Repair of crosslinked DNA and survival of *Escherichia coli* treated with psoralen and light: effects of mutations influencing genetic recombination and DNA metabolism, *J. Bacteriol.,* 136, 538, 1978.

68. **Lin, P. F., Bardwell, E., and Howard-Flanders, P.,** Initiation of genetic exchanges in lambda phage-prophage crosses, *Proc. Natl. Acad. Sci. U.S.A.,* 74, 291, 1977.

69. **Cassutto, E., Gross, N., Bardwell, E., and Howard-Flanders, P.,** Genetic effects of photoadducts and photocrosslinks in the DNA of phage lambda exposed to 360 nm light and trimethylpsoralen or khellin, *Biochim. Biophys. Acta,* 475, 589, 1977.

70. **Ross, P. and Howard-Flanders, P.,** Initiation of recA+-dependent recombination in *Escherichia coli* (lambda). I. Undamaged covalent circular lambda DNA molecules in uvrA+recA+ lysogenic host cells are cut following superinfection with psoralen-damaged lambda phages, *J. Mol. Biol.,* 117, 137, 1977.

71. **Ross, P. and Howard-Flanders, P.,** Initiation of recA+-dependent recombination in *Escherichia coli* (lambda). II. Specificity in the induction of recombination and strand cutting in undamaged covalent circular bacteriophage 186 and lambda DNA molecules in phage-infected cells, *J. Mol. Biol.,* 117, 159, 1977.

72. **Cupido, M. and Bridges, B. A.,** Uvr-independent repair of 8-methoxypsoralen crosslinks in *Escherichia coli:* evidence for a recombinational process, *Mutat. Res.,* 146, 135, 1985.

73. **Hall, J. D. and Scherer, K.,** Repair of psoralen-treated DNA by genetic recombination in human cells infected with herpes simplex virus, *Cancer Res.,* 41, 5033, 1981.

74. **Hall, J. D.,** Repair of psoralen-induced crosslinks in cells multiply infected with SV40, *Mol. Gen. Genet.,* 188, 135, 1982.

75. **Jachymczyk, W. J., von Borstel, R. C., Mowat, M. R. A., and Hastings, P. J.,** Repair of interstrand cross-links in DNA of *Saccharomyces cerevisiae* requires two systems for DNA repair: the RAD3 system and the RAD51 system, *Mol. Gen. Genet.,* 182, 196, 1981.

76. **Magana-Schwencke, N., Henriques, J.-A. P., Chanet, R., and Moustacchi, E.,** The fate of 8-methoxypsoralen photoinduced crosslinks in nuclear and mitochondrial yeast DNA: comparison of wild-type and repair-deficient strains, *Proc. Natl. Acad. Sci. U.S.A.,* 79, 1722, 1982.

77. **Saeki, T., Cassier, C., and Moustacchi, E.,** Induction in *Saccharomyces cerevisiae* of mitotic recombination by mono and bifunctional agents: comparison of the pso2-1 and rad52 repair deficient mutants to the wild-type, *Mol. Gen. Genet.,* 190, 255, 1983.

78. **Carter, D. M., Wolff, K., and Schnedl, W.,** 8-Methoxypsoralen and UVA promote sister-chromatid exchanges, *J. Invest. Dermatol.,* 67, 548, 1976.

79. **Waksvik, H., Brogger, A., and Stene, J.,** Psoralen/UVA treatment and chromosomes. I. Aberrations and sister chromatid exchange in human lymphocytes in vitro and synergism with caffeine, *Hum. Genet.,* 38, 195, 1977.

80. **Mourelatos, D., Faed, M. J. W., and Johnson, B. E.,** Sister chromatid exchanges in human lymphocytes exposed to 8-methoxypsoralen and long wave UV radiation prior to incorporation of bromodeoxyuridine, *Experientia,* 33, 1091, 1977.

81. **Wulf, H. C.,** Acute effect of 8-methoxypsoralen and ultraviolet light on sister chromatid exchange, *Arch. Dermatol. Res.,* 263, 37, 1978.

82. **Gaymor, A. L. and Carter, D. M.,** Greater promotion in sister chromatid exchanges by trimethylpsoralen than by 8-methoxypsoralen in the presence of UV-light, *J. Invest. Dermatol.,* 71 257, 1978.

83. **Carter, D. M., Lyons, M. F., and Windhorst, D. B.,** Photopromotion of sister chromatid exchanges by psoralen derivatives, *Arch. Dermatol. Res.,* 272, 239, 1982.

84. **West, M. R., Johansen, M., and Faed, M. J. W.,** Sister chromatid exchange frequency in human epidermal cells in culture treated with 8-methoxypsoralen and long-wave UV radiation, *J. Invest. Dermtol.,* 78, 67, 1982.

85. **Latt, S. A. and Loveday, K. S.,** Characterization of sister chromatid exchange induction by 8-methoxypsoralen plus near UV light, *Cytogenet. Cell Genet.,* 21, 184, 1978.

86. **Cassel, D. M. and Latt, S. A.,** Relationship between DNA adduct formation and sister chromatid exchange induction by [³H]8-methoxypsoralen in Chinese hamster ovary cells, *Exp. Cell Res.,* 128, 15, 1980.

87. **Shuler, C. F. and Latt, S. A.,** Sister chromatid exchange induction resulting from systemic, topical, and systemic-topical presentations of carcinogens, *Cancer Res.,* 39, 2510, 1979.

88. **Bredberg, A., Lambert, B., Lindblad, A., Swanbeck, G., and Wennersten, G.,** Studies of DNA and chromosome damage in skin fibroblasts and blood lymphocytes from psoriasis patients treated with 8-methoxypsoralen and UVA irradiation, *J. Invest. Dermatol.,* 81, 93, 1983.

89. **Swanbeck, G., Thyresson-Hok, M., Bredberg, A., and Lambert, B.,** Treatment of psoriasis with oral psoralens and longwave ultraviolet light, *Acta Derm. Venereol. (Stockholm),* 55, 367, 1975.

90. **Wolff-Schreiner, E. C., Carter, D. M., Schwarzacher, H. G., and Wolff, K.,** Sister chromatid exchanges in photochemotherapy, *J. Invest. Dermatol.,* 69, 387, 1977.

91. **Brogger, A., Waksvik, H., and Thune, P.,** Psoralen/UVA treatment and chromosomes. II. Analyses of psoriasis patients, *Arch. Dermatol. Res.,* 261, 287, 1978.

92. **Sahar, E., Kittrel, C., Fulghum, S., Feld, M., and Latt, S. A.,** Sister-chromatid exchange induction in Chinese hamster ovary cells by 8-methoxypsoralen and brief pulses of laser light. Assessment of the relative importance of 8-methoxypsoralen-DNA monoadducts and crosslinks, *Mutat. Res.,* 83, 91, 1981.

93. **Bredberg, A. and Lambert, B.,** Induction of SCE by DNA crosslinks in human fibroblasts exposed to 8-MOP and UVA irradiation, *Mutat. Res.,* 118, 191, 1983.

94. **Liu-Lee, V. W., Heddle, J. A., Arlett, C. F., and Broughton, B.,** Genetic effects of specific DNA lesions in mammalian cells, *Mutat. Res.,* 127, 139, 1984.

95. **Natarajan, A. T., Verdegaal-Immerzeel, E. A. M., Ashwood-Smith, M. J., and Poulton, G. A.,** Chromosomal damage induced by furocoumarins and UVA in hamster and human cells including cells from patients with ataxia telangiectasia and xeroderma pigmentosum, *Mutat. Res.,* 84, 113, 1984.

96. **Linnainmaa, K. and Wolff, S.,** Sister chromatid exchange induced by short-lived monoadducts produced by the bifunctional agents mitomycin C and 8-methoxypsoralen, *Environ. Mutat.,* 4, 239, 1982.

97. **Loveday, K. S. and Donahue, B. A.,** Induction of sister chromatid exchanges and gene mutations in Chinese hamster ovary cells by psoralens, *Natl. Cancer Inst. Monogr.,* 66, 149, 1984.

98. **Billardon, C., Levy, S., and Moustacchi, E.,** Induction in human skin fibroblasts of sister-chromatid exchanges by photoaddition of two new monofunctional pyridopsoralens in comparison to 3-carbethoxypsoralen and 8-methoxypsoralen, *Mutat. Res.,* 138, 63, 1984.

99. **Vijayalaxmi, and Wunder, E.,** Comparison of sister chromatid exchange induced by photoactivated 3-carbethoxypsoralen and 8-methoxypsoralen in human blood lymphocytes, *Mutat. Res.*, 152, 211, 1985.

100. **Ashwood-Smith, M. J., Grant, E. L., Heddle, J. A., and Friedman, G. B.,** Chromosome damage in Chinese hamster cells sensitized to near-ultraviolet light by psoralen and angelicin, *Mutat. Res.*, 43, 377, 1977.

101. **Hook, G. J., Heddle, J. A., and Marshall, R. R.,** On the types of chromosomal aberrations induced by 8-methoxypsoralen, *Cytogenet. Cell Genet.*, 35, 100, 1983.

102. **Miller, S. S. and Eisenstadt, E.,** Suppressible base substitution mutations induced by angelcin in the *E. coli* lac I gene: implications for the mechanism of sos mutagenesis, *J. Bacteriol.*, 169, 2724, 1987.

Chapter 7

REPAIR OF DNA CONTAINING FUROCOUMARIN ADDUCTS

Charles Allen Smith

TABLE OF CONTENTS

I. INTRODUCTION

Interest in cellular processing of furocoumarin adducts in DNA derives from a number of sources. The use of furocoumarins in photochemotherapy (see Chapter 8) and as tanning aids,[1] and their presence at possibly genotoxic levels in certain foods[2,3] provide compelling reasons for understanding their carcinogenic potential. Furocoumarin adducts also provide an attractive model system to study basic DNA repair mechanisms for several reasons:

1. The requirement for photoactivation to produce adducts facilitates control of their introduction to a degree not usually available with other reactive chemicals.
2. The adducts are stable to most chemical procedures and are not spontaneously lost from the DNA in vivo.
3. Although they exist in various isomeric forms, all the adducts involve a cyclobutane bridge to the 5,6 carbons of a pyrimidine base, usually thymine. This is the same structure involved in the well-characterized cyclobutane dipyrimidine formed by short-wave UV radiation.
4. The interstrand cross-link that can be generated by many furocoumarins presents obstacles to DNA transactions fundamentally different from those imposed by most DNA lesions that involve only one strand. Not only is information compromised on both strands simultaneously, but semiconservative DNA replication cannot proceed until the linkage is broken. The ability to vary the ratio of cross-links to monoadducts at constant total adduct frequencies, the availability of some furocoumarins that yield only monoadducts, and the stability of the adducts are attributes not shared by other cross-linking agents.
5. The retention of cross-linking potential by many of the monoadducts in purified DNA facilitates analysis of their fate using sensitive techniques based upon special properties of cross-linked DNA.

The introduction and processing of psoralen adducts in DNA have been studied by many investigators, using a wide range of organisms and cultured cells. Discussion of this topic may be found in a number of recent reviews concerning furocoumarins[4-7] or particular DNA-damaging agents.[8-9] Reviews on DNA repair include References 10 through 18.

The present chapter will focus on mechanisms of repair of these adducts and how the particular features of furocoumarins may be used to further our understanding of processing of lesions by cells. Ultimately one would like to know the molecular mechanisms for recognizing and removing the adducts, how the operation of these mechanisms is influenced by the location of a given adduct, the overall system capacity and kinetics for removal of adducts, and the consequences of adduct processing for DNA integrity and various biological endpoints.

For simplicity, throughout this discussion, it should be assumed that statements of the type "treated with psoralen" indicate treatment with psoralen plus irradiation with long-wavelength UV light (UVA). Most investigations have included controls to demonstrate that the effects under study require both treatments. The term UV will refer to 254-nm light.

II. RELEVANT DNA REPAIR MECHANISMS

A brief review of the known mechanisms for processing DNA damage is necessary before focusing on psoralen adducts in particular. Although pathways for repair are often described as if they are mutually exclusive, this is not the case, as will become apparent when we consider models for the repair of cross-linked DNA. A given adduct may be processed by different pathways under different circumstances.

A. Excision Repair

This mechanism operates in most cells to remove lesions in DNA. For damage in only one of the two strands, the process is simple in concept: a segment of DNA containing the damage is removed and replaced by new DNA synthesis, using the complementary strand as template. The details of the enzymatic mechanisms can vary according to the organism or cell type, the lesion being repaired, and possibly the function or activity of the damaged DNA sequence.

Several distinctly different mechanisms have been identified that initiate the DNA strand breakage required for removal of the damaged segment. A damaged base may be removed by a DNA glycosylase, leaving an AP site that is subject to the action of one or more AP endonucleases, which cleave phosphodiester bonds at such sites. With a few exceptions each of the known glycosylases acts only on a specific inappropriate or damaged base (e.g., uracil or methylated purines). The well-studied pyrimidine dimer-specific endonucleases from *M. luteus* and phage T4-infected *Escherichia coli* are in fact dimer-specific glycosylases with associated AP endonuclease activities. Another mechanism for incision displays a broad specificity, acting on a variety of bulky addition products and other DNA-distorting lesions such as pyrimidine dimers. Genetic evidence indicates that in both prokaryotes and eukaryotes this mechanism utilizes several proteins acting in concert to recognize the damage and incise the DNA. The *E. coli* UvrABC nuclease, reconstituted in vitro from the purified products of the *uvrA, B,* and *C* genes, can cut a damaged strand on both sides of the lesion, with each incision several nucleotides away from the lesion itself. Subsequent to incision, the concerted action of exonucleases, DNA polymerases, and helicases replace a stretch of nucleotides using the complementary strand as template. The repair patch is then made contiguous to the parental strand by polynucleotide ligase. The extent of new synthesis may depend on the incision mechanism, but with the exception of the inducible long-patch excision repair (ER) observed in *E. coli*,[19] less than about 50 nucleotides are replaced. ER has been studied extensively in bacteria, yeast, and mammalian cells, and it is considered to be error free, at least when the lesions are widely spaced in the DNA.

B. Mismatch Repair

Mismatched bases arise predominately from the insertion of incorrect nucleotides during replication, although the deamination of cytosine and 5-methyl cytosine in DNA to create uracil and thymine also occurs. Details of the molecular mechanism of mismatch repair in bacteria are being elucidated, but not much is known for other organisms. The need for a means to determine which base is the incorrect one is crucial in such a system; *E. coli* uses transient undermethylation of the newly replicated DNA for repair of mismatches immediately behind the growing fork. In other regions of the DNA or in organisms which do not use a tag such as methylation, the required information for mismatch correction may come from a second duplex copy of the same sequence. In a formal sense, interstrand cross-links could be repaired by such a mismatch correction mechanism.

C. Repair of Daughter Strand Gaps

This process, also known as postreplication repair or recombinational repair, operates on gaps left opposite lesions in parental DNA as the replication fork traverses the damaged region. The gap may be filled using an isopolar copy of the sequence in an error-free recombinational process or by generally error-prone synthesis over the noninstructional site by a polymerase. The former mechanism has been studied in detail in *E. coli*, and the latter has been proposed as a major route for mutagenesis. Before such a gap is filled, the operation of ER on the lesion would generate a double-strand break.

D. Repair of Double-Strand Breaks

The repair of double-strand breaks such as those caused by ionizing radiation has been demonstrated in several kinds of cells. Double-strand breaks have also been observed in cells undergoing repair of other damage, and models have been presented for their repair, generally requiring the presence of an intact homologous copy of the affected region.[20,21]

E. Possible Models for Repair of Furocoumarin-Damaged DNA

A priori, psoralen monoadducts appear to be good candidates for the repair systems that process other bulky lesions like pyrimidine dimers. Cross-links, on the other hand, present unique problems because information is compromised on both DNA strands at essentially the same site. However, because the two affected bases are staggered it is possible to envision schemes by which the base immediately opposite one of the bases involved in the cross-link provides the information for the proper replacement of that particular base, leaving damage only on one strand to be processed by ER. The simplest of such models invokes a DNA glycosylase to remove one or both bases involved in the cross-link, followed by insertion of proper bases either in some direct manner or by the sequential action of AP endonuclease(s) to leave a single nucleotide gap to be filled by polymerization. In the absence of a glycosylase mechanism, the schemes that can be drawn using endonucleases become complex enough to appear to be rather unlikely. In some configurations the intercalated psoralen molecule is on the 5' side of each involved base, while in others it is on the 3' side. To avoid using the noninstructional site on the template strand for repair synthesis, a damage-specific endonuclease would have to make an incision on the same side of the base as the intercalated psoralen (5' in some cases and 3' in others). The mechanism of exonuclease action and the direction of polymerase action (away from or toward the damage site) must also differ for the two configurations of cross-links. Moreover, in one of the two configurations the ligation would have to occur only one nucleotide away from the remaining damaged base in the opposite strand. A more plausible scheme for cross-link repair, and one for which there is experimental support, uses recombination with a second copy of the relevant sequence present in the cell to provide the information to replace one of the strands, rather than attempting to use its complement.

Schemes for removing the cross-link that do not restore the original nucleotide sequence are easier to envision and may contribute to mutagenesis directed by these lesions. Repair synthesis using the still-damaged template could occur, or some process might excise a portion of both strands containing the cross-link, eventually generating a deletion while repairing the consequent double-strand break.

III. METHODS FOR MEASURING REPAIR

Repair of DNA can be monitored by a number of different techniques, ranging from direct determination of frequencies of chemically defined adducts to examination of gross changes in chromatin by microscopic observation.

A. Direct Measurements of Adduct Frequency

Measurements using HPLC are described in Chapter 2. The major psoralen adducts can also be resolved by continuous flow gel electrophoresis.[22,23] The sensitivity of high performance liquid chromatography (HPLC) and electrophoresis is achieved by using radiolabeled psoralen or prelabeling the cellular DNA, and it is limited by specific activities obtainable and the amount of material that can be subjected to analysis. Frequencies in the range of 0.5 adducts per 100 kbases (kb) are easily measured. However, most determinations of adduct frequencies to date have come from measurements of specific activity of DNA purified from cells treated with radioactive furocoumarins, giving only the frequency of total adducts. Until recently, radiolabeled furocoumarins that yield only monoadducts have not been generally available, and so most determinations of adduct frequencies have been studied with compounds that make both monoadducts and cross-links.

B. Indirect Measurements of Adduct Frequency

The immunological methods described in Chapter 2 appear to hold great promise but have not been extensively used in studies of repair. Another method yet to be fully explored would use *E. coli* UvrABC nuclease to break DNA at the sites of psoralen adducts in much the same way that pyrimidine dimer-specific nucleases have been used to quantitate those lesions.[24]

An indirect method that has been used extensively to determine lesion frequency is the cross-linking assay. Nearly all of the numerous forms of this assay separate single-stranded from duplex DNA after the DNA is briefly subjected to denaturing conditions. Only those molecules containing a cross-link rapidly renature. Methods for distinguishing single-stranded from duplex DNA have included chromatography on hydroxylapatite or BND cellulose, centrifugation in isopycnic density gradients (optimally at a pH just below the denaturation point to maximize the density difference between the two forms), gel electrophoresis (applicable only to DNA molecules or fragments of uniform size), and treatment with single-strand-specific nucleases to degrade all but the duplex DNA. Velocity centrifugation under denaturing conditions can be used to measure the fraction of cross-linked molecules if they are of uniform size, and analysis in alkaline CsCl density gradients has been used to measure cross-linking between normal and BrdUrd-substituted DNA strands. Alkaline elution can also be used to determine DNA cross-linking if appropriate controls are included.

The cross-linking assay can be combined with reirradiation of DNA containing psoralen monoadducts after removal of unbound psoralen (often after DNA purification) as an indirect method to measure those monoadducts capable of forming cross-links. This requires knowing the quantum yield for conversion of monoadduct to cross-links and the maximum conversion possible in DNA.

It should be kept in mind that the loss of cross-linking does not necessarily imply removal of the adducts from one or both strands. Nicking of the DNA at the site of the lesion (especially a nick on each side), while hardly constituting repair, would appear as a loss of cross-linking. Extensive random nicking could have the same effect.

C. Indirect Measurements of Repair

Many methods are available to study repair that do not measure adducts directly. The resynthesis step in ER can be monitored by measuring the incorporation of radiolabeled precursor into the DNA. For growing cells, stretches of new DNA arising from normal semiconservative replication can be removed from the analysis by density labeling. This method is very sensitive at early stages in repair, when few adducts have been removed and adduct loss cannot be accurately measured directly. It is simple and quantitative, and can be used to compare repair with various damaging agents, doses, or treatment conditions, or to make comparisons of different cell types or mutants. It can also be used to determine the size of the resynthesized tract.[25] A less quantitative measurement usually termed unscheduled DNA synthesis (UDS) may be obtained by autoradiographic detection of incorporated precursor. This method, applied mostly to mammalian cells, has been particularly useful for classifying ER-deficient mutants into complementation groups. It may also be used to study population heterogeneity.

Damage-dependent incision of cellular DNA can be monitored by assaying the frequency of single-strand breaks in the DNA of cells undergoing repair. This can be done by including polymerase inhibitors that prevent the rapid sealing of such breaks upon the completion of repair or by using ligase-deficient mutants.

IV. REPAIR IN *ESCHERICHIA COLI*

The ease with which both biochemical and genetic studies can be carried out and correlated in this organism has led to an understanding of its repair processes at a level of detail not yet approached with eukaryotes. Many of the enzymes involved in repair have been characterized and their genes identified. Complex coordinated response networks induced by damage, such as those responding to UV and other bulky adducts, alkylating agents, oxidative damage, and heat, are becoming understood. The most important for the present discussion is the regulon under control of the *recA* and *lexA* genes (the so-called SOS system). Because our knowledge of the biochemical basis underlying the phenotypes of many repair-deficient mutants has advanced so much in the past few years, some of the conclusions drawn from previous investigations of repair of furocoumarin damage require careful reexamination. For example, before the role of the *recA* gene product in regulating repair responses was appreciated, there was a tendency to view excision repair and recombinational repair as independent processes, controlled respectively by the *uvr* genes and *rec* genes. We now know that *recA* mutations can exert effects on repair phenomena both by altering the induction of gene products involved in repair and by altering other activities of the *recA* protein itself.

A. Repair of Monoadducts

This can be done either with furocoumarins like angelicin and 3-carbethoxypsoralen (3-CPs) that make only monoadducts, or with psoralens under irradiation conditions that produce few if any cross-links. Bordin et al.[26] observed 80% removal of radiolabeled angelicin by repair-proficient *E. coli* in 30 min at an initial frequency of about one monoadduct per 100 kb. Since no survival data were presented, it is not possible to compare the biological responses to these adducts in a quantitative way to those for UV-induced pyrimidine dimers. Grossweiner and Smith[27] showed that *recA*, *uvrB*, or *polA1* mutants of *E. coli* K-12 are hypersensitive to angelicin, a situation smilar to that for short-wavelength UV radiation. Bridges et al.[28] concluded that the inactivation of *uvrA recA* mutants of *E. coli* B by 8-methoxypsoralen (8-MOP) is a one-photon process with a quantum yield of about 0.03, and therefore that, as with pyrimidine dimers, one or a very few monoadducts is lethal to such repair-deficient mutants. Cole had originally concluded that monoadducts were relatively innocuous in K-12 derivatives,[29] but later estimates of the frequency of adducts produced under Cole's conditions[30] and an appreciation of the differences in photoproduct yield for different derivatives probably account for the apparent contradiction. Something in K-12 hinders interaction with the DNA, producing much lower yields than in B strains, even though intracellular concentrations of 8-MOP are similar.[31]

These observations suggest that psoralen monoadducts are processed in wild-type bacteria in the same manner as are other bulky adducts, such as pyrimidine dimers. The demonstration that the UvrABC nuclease acts upon monoadducts[32,33] in pure DNA in a manner similar to that on DNA containing pyrimidine dimers is consistent with this idea, as are the results of studies on the inactivation of phage. T3 phage treated with 8-MOP under conditions that produce few cross-links survive better on wild-type hosts than on *uvrA* mutants. In the case of phage T4, the bacterial incision system does not appear to act on the phage DNA, and the T4 *denV* gene product (endonuclease V) is specific for pyrimidine dimers. Thus, mutations in the *denV* gene do not affect resistance to 8-MOP monoadducts[34] or angelicin adducts.[35] The T4 postreplication repair system coded by the *uvsW*, *uvsX*, and *uvsY* genes does appear to act on phage genomes containing angelicin.

Bridges and Stannard[36] presented evidence that *uvrA* mutants can process cross-linkable monoadducts in such a way that reduces the lethal effect of converting them to cross-links by further UVA irradiation after removal of unbound psoralen. This processing did not require growth conditions, but was inoperative at low temperature or in a *polA* mutant, which suggests that an active repair process is involved rather than some spontaneous alteration of the monoadduct. The assay used was necessarily indirect and provided no indication of the number of adducts involved. In light of the possibility that some residual activity of the UvrABC nuclease accounts for these results, the experiments deserve to be repeated in *uvrB* deletion mutants or strains carrying multiple *uvr* mutations.

Recently, *uvrB umuC* strains have been shown to be markedly more sensitive to angelicin than are the corresponding *uvrB umuC*⁺ strains.[37] This is not the case for UV damage and suggests that, at least in excision-deficient strains, the *umuC* gene product, necessary for UV mutagenesis, facilitates replication past angelicin monoadducts.

Daughter-strand gap repair probably operates efficiently on gaps opposite monoadducts, but no studies have been presented to document this.

B. Repair of Cross-links

In a series of investigations over 8 years, Cole and his colleagues carried out early biochemical studies concerned with repair of interstrand cross-links, leading to a model for their removal.[29,30,38-43] A moderately detailed account of the model and its support is warranted here as it has been extensively used to guide studies with other organisms.

Cole and his collaborators were able to resolve the process of cross-link repair into two distinct phases and to study them independently using various mutants deficient in repair or recombination. Immediately after exposure of cells to trimethylpsoralen (TMP) and UVA, their DNA sedimented more rapidly in alkaline sucrose gradients than did DNA from control cells. After 30 min post-treatment incubation, the DNA sedimented much more slowly, with a number average molecular weight corresponding to about twice the average inter-cross-link distance. Over the next 90 min the sedimentation rate increased with incubation time. The initial lowering in molecular weight was not observed in *uvrA* or *uvrB* mutants, and the subsequent increase in size of DNA was absent in *recA* mutants. These data suggested a model in which incisions are first placed on either side of each cross-link in one of the two strands of the DNA (chosen at random at each site), followed by recombinational strand exchanges involving a second copy of the chromosome to restore the proper DNA sequence between the two incisions. After this, normal ER could remove the residual damage to the other strand and replace the missing nucleotides using the newly imported DNA strand as template.

Removal of DNA cross-linking was then shown to occur with the same kinetics as the reduction of apparent molecular weight and to be dependent on the *uvrA, B,* and *C* gene products. Interestingly, the loss of cross-linking was reduced in two mutants defective in the 5'-3' exonuclease activity of polymerase I (*polA107* and *polAex4*), but not in mutants deficient in the polymerizing activity. Further studies showed that the molecular weight changes observed using sucrose gradients and the loss of cross-linking measured as rapidly renaturing DNA exhibited the same genetic control, and that double-strand breaks did not occur to any measurable extent during the incision process. Density labeling experiments also supported the hypothesis that cuts were made in one of the two strands, on either side of the cross-link.

At the time of these studies the incision activity coded by the *uvrA, B,* and *C* genes was thought to make a single cut 5' to a damaged site in DNA. The dependence of loss of cross-linking on polI exonuclease mutants and the properties of the nuclease activity in vitro strongly suggested that it was responsible for making incisions 3' to the cross-links. However, recent studies with the purified UvrABC nuclease[44,45] have shown that the complex is capable of making incisions on both sides of a cross-linked nucleotide, specifically cutting the strand linked to the furan end of the molecule. This suggests some other role for the polI 5'-3' exonuclease activity in incising cross-linked DNA. In vitro, the proteins that make up the UvrABC nuclease remain bound to the substrate DNA after making incisions at pyrimidine dimers. Catalytic action of the incision activity is thus dependent upon release of the proteins, which can be achieved in vitro with chelating agents or the addition of both the *uvrD* product (helicase II) and DNA polymerase I under polymerizing conditions.[46] Assuming that the complex can make dual incisions in vivo, the exonuclease function of polI might be necessary for release of the complex to allow it to make incisions at all damage sites. It is interesting in this regard that Yoakum and Cole[41] observed deficient incision at cross-links in toluene-permeabilized *uvrD* mutants. This result was attributed at the time to a role for the *uvrD* product in regulating production of the second incision at cross-links, but could now be interpreted as a reflection of limited overall incision in these mutants.

Since most conditions produce far more monoadducts than cross-links, the possibility should be kept in mind that deficiencies in monoadduct processing could have indirect effects on cross-link processing. More detailed studies of effects of other proteins on the activity of UvrABC nuclease at monoadducts and cross-links in vitro should help explain the results obtained with various mutants in vivo.

The postincision events in cross-link removal are still poorly understood. Early density labeling experiments provided evidence to support *recA*-dependent exchanges of relatively long (>3000 nucleotides) stretches of DNA. The strand rejoining process was measured in a large number of mutants deficient in various steps in recombination, using sucrose gradient assays. All of these mutants were more sensitive to the killing action of the psoralen adducts than were wild-type cells, and all were found to be defective in the rejoining process to some degree. Both the RecBC and RecF pathways appeared to participate in the rejoining process. The *lexA3* mutant exhibited no rejoining, suggesting a requirement for induction of increased amounts of RecA protein and/or induction of some other gene(s) controlled by the *recA-lexA* regulon. The *lexA* gene normally produces a repressor that is involved in control of a number of damage inducible genes, including *recA*. Cleavage of the *lexA* gene product is facilitated by activated RecA protein. However, temperature shift experiments with the thermosensitive mutant *recA200* indicated that functional RecA protein was needed during the entire rejoining process. This suggests RecA protein also participates directly in the strand exchange.

During rejoining, high-molecular-weight DNA was observed to increase in amount at the expense of fully incised DNA, instead of by a pattern characterized by a gradual shift in the molecular weight of all the DNA. This suggested that the rejoining occurred in clusters (perhaps representing supercoiled loops in the chromosome) or started at a fixed point and proceeded progressively away from it.

Rejoining occurred at a reduced rate in *polA1* mutants, but was eliminated in *polA polB dnaE*[ts] strains at 42°, but not at 32°, suggesting that the process normally uses DNA polymerase I for the DNA synthesis associated with the exchange process, but that DNA polymerase III can substitute in its absence. The *polA* and *polB* genes code for DNA polymerases I and II respectively, and *dnaE* codes for DNA polymerase III, which is necessary for normal chromosomal DNA replication.

The events that lead from incision to recombination are not at all clear. In vitro, the UvrABC nuclease leaves a 3'-OH terminus eight nucleotides from the site of the cross-link.[44] It would be reasonable to assume that polymerase I could carry out nick translation only up to the cross-link. If a gap of about 1000 nucleotides is opened at the 3' side of the incision,[43] the resultant structure would be formally analogous to that formed when a replication fork stalls at a lesion and replication is reinitiated at the next Okasaki start site. This structure is thought to be the substrate for daughter strand gap repair, which also uses strand exchanges to fill in the gap, probably using a stretch of sister duplex formed by replication after the damage. However, in the case of cross-link repair, either a second copy of the chromosome or a sister duplex formed by replication *before* damage was introduced must be involved, since the damage involves both strands at nearly the same point on the duplex.

Grover et al.[47] showed that the survival of TMP-treated cells was markedly lower when growing slowly (assumed to contain only one chromosome) than under conditions of rapid growth (containing several). However, when attempts were made to model the system taking into account the amount of two-copy DNA due to replication and the time delay before cell division, more cross-link repair (judged by survival) was observed than was expected if two copies were necessary for repair. It appears that biochemical studies on some suitably synchronized cell system would be necessary to settle the point; however, forcing cells into such an artificial situation may alter their capacity for repair in other ways, e.g., in their capacity to induce SOS. Another approach to determining the necessity of two copies of the chromosome is to study repair of exogenously supplied DNA, such as plasmid DNA, in which each molecule contains a cross-link at the same site. In a recent study[48] cross-links were placed specifically at one site in a polylinker DNA fragment, and cross-linked fragments were purified and ligated upstream of the *lacZ* gene in a plasmid containing a selectable antibiotic resistance marker. Pretreatment of the cells with UV greatly increased the number of transformants obtained with cross-linked plasmid; presumably introduction of the damaged plasmid was insufficient to induce some SOS functions required for the repair. Surprisingly, 97 out of 100 transformed colonies harbored plasmid containing the restriction site into which the psoralen had been introduced. This result strongly suggests an inducible error-free mechanism for removal of cross-links that does not utilize a second copy of the information compromised by the cross-link. More studies of this type, especially using various repair-deficient mutants, will be needed to evaluate the efficiency of this mode of repair and its relation to repair in the chromosome.

A number of other studies have also shown relatively rapid removal of radiolabeled furocoumarins or cross-linking, and demonstrated involvement of the *uvr* and *rec* genes similar to that already discussed.[26,27,49,50]

Bridges and his co-workers have studied in some detail the limited amount of cross-link removal that occurs in cells deficient in UvrABC nuclease, relying mainly on survival assays to estimate repair.[51-54] Such repair is greatly reduced in *recA* mutants, facilitated in cells containing functional helicase II (the *rep* gene product), and inhibited by acriflavine. Although the repair was dependent to some extent on the *umuC* product, it was inhibited in the presence of the product of the *muc* operon of plasmid pKM101. This is unexpected since the *muc* operon is analogous in its effect on UV-induced mutagenesis to the endogenous *umuC* gene in *E. coli*. This repair pathway can be considered independent of even residual UvrABC nuclease activity because of its appearance in *uvrB* deletion mutants, but *uvrA* insertion mutants showed less inhibition of repair by acriflavine. The pathway appears to be part of the SOS system as judged by the fact that cells in which SOS induction is brought about by a temperature shift *(recA441)* survive better than such cells not induced. This scheme also appears to require recombination with a second copy of the genome, using either the RecBC or RecF pathway. Cells containing only a single chromosome were markedly more sensitive than cells with many chromosomes, and (under SOS-induced conditions) survival of cross-linked lambda phage indicated repair of cross-links only when the multiplicity of infection was greater than one. Models to account for this *uvrA* independent repair have changed over the course of these studies. The critical question concerns the mechanism by which the DNA is incised at cross-links. A glycosylase activity was originally proposed, but it is hard to understand why such an activity would act only at DNA unwound by the Rep helicase and why it would not also act at furocoumarin monoadducts as well as cross-links. The most recent report suggests the double-strand breaks and their repair via a recombinational process might be involved.[54]

In summary, psoralen monoadducts in wild-type *E. coli* appear to be treated very similarly to a number of other bulky adducts, with possibly some minor differences in the role of *umuC* gene product at persisting adducts. The major pathway for elimination of psoralen cross-links is initiated by incision on both sides of the cross-link in one of the strands by UvrABC nuclease, followed by some processing events at the incision sites that presumably lead to a well-coordinated sequence of recombinational events with a homologous copy of the relevant sequence. Functional RecA protein, probably in increased amounts, is necessary for the recombination, and other damage-inducible genes may also be required. In the absence of UvrABC nuclease, the processing of a small number of both monoadducts and cross-links results in their removal or alteration to less lethal forms. The mechanism for this pathway is poorly understood, but it might occur at some adducts in repair-proficient cells as well.

Much remains to be explored, especially in the area of the postincision processing. The nature of the well-documented involvement of the SOS system in the processing of cross-links can now be determined much more specifically by examining the biochemistry of repair in mutants with specific defects in this response. Simple assays of survival are inadequate because the SOS response is important for resistance to agents like UV that make damage to only one strand;[55] one would expect the lethality of psoralen monoadducts to be ameliorated by these processes as well. Recently, *E. coli* mutants hyperresistant to 8-MOP were described and partially characterized.[56] A 55-kdalton protein was shown to be over-expressed in these mutants, but not in *recA* derivatives of them. These derivatives also did not exhibit increased resistance relative to *recA* derivatives of the parent cells. Thus the enhanced resistance appears unrelated to photoproduct yield (as with the strain differences noted above or in *acr* mutants[57]) and may lead to identification of a damage-inducible protein involved in cross-link repair.

V. REPAIR IN *SACCHAROMYCES CEREVISIAE*

Repair in this single-celled eukaryotic organism has been studied for some time using genetic techniques. Genetic evidence indicates a complexity in repair pathways far exceeding that understood at the biochemical level. Recent developments in the ability to do biochemistry and genetic engineering in yeast make it likely that it will in time rival *E. coli* as an organism from which models for repair in higher eukaryotes are built.

Genetic evidence for different repair pathways in yeast has been obtained mainly by study of the survival of strains containing combinations of mutations conferring sensitivity to different DNA-damaging agents.[12,13,58,59] When a double mutant exhibits no greater sensitivity than the more sensitive of the single mutants, the two mutations are termed epistatic, and it is assumed that the mutant genes control steps in the same pathway for repair. Three different epistasis groups have been identified using UV and ionizing radiation as damaging agents. The groups are rather large and are designated by the name of a prominent locus in each. If the sensitivity of the double mutant appears to be the sum of the sensitivities of the single mutants, it is inferred that the genes are involved in pathways that repair different lesions in the DNA. When the sensitivity of the double mutant exceeds the sums of the sensitivities of the single mutants (a synergistic effect), it is thought that the pathways involved represent different mechanisms for processing the same lesion(s), but that neither pathway alone can effect complete repair. These interpretations are by no means straightforward (discussed in References 58 and 59), especially since survival may vary with the growth conditions (exponential or stationary) and ploidy of the cells. However, other properties common to mutants in the different epistasis groups can also help to define the actual mechanisms involved in the pathways themselves. Some mutants have been classified according to these properties, even if their assignment from sensitivity studies is unclear.

The *RAD3* epistasis group (at least ten members) appears to correspond to the *uvr* mutants in *E. coli* and to mediate ER of bulky lesions. They are sensitive to UV and other agents that produce bulky adducts; several methods have been used to show that they are generally defective in removing pyrimidine dimers. Five of the mutants appear to be unable to perform incision, and another five exhibit partial deficiency in removal of pyrimidine dimers. Considerable progress has been made in cloning these genes, and perhaps we will soon have a description of the mechanism of incision mediated by their products.

The *RAD52* group (at least ten members, including *rad50* to *rad57*) is characterized by sensitivity to ionizing radiation and is only slightly sensitive to UV. These mutants are defective in both induced and spontaneous recombination, and several of them fail to repair double-strand breaks. A recent model in which recombination in yeast is initiated by a double-strand break is consistent with the notion that double-strand break repair and recombination share enzymatic steps.[21] At present there seems to be no evidence for regulation of damage-inducible genes by proteins also involved in recombination, as is the case in *E. coli*.

The mutants in the *RAD6* group (at least 14 members) are sensitive to both UV and ionizing radiation, but show no mutagenesis induced by UV, certain chemicals, or ionizing radiation. This suggests that they mediate replication past noninstructional sites, such as in gaps formed opposite lesions in replicating DNA. The term mutagenic or error-prone repair is often applied to the process deficient in the members of the group, although use of the term repair in this context is dubious. Genes in this group may also participate in a minor way in excision repair and recombinational repair.

A. Repair of Monoadducts

As with *E. coli*, repair of angelicin adducts in yeast appears to resemble that of pyrimidine dimers. Excision repair mutants (*RAD3* group) are sensitive to angelicin,[60,61] but *rad51* mutants (belonging to the *RAD52* group) are sensitive only at high doses. Transient single-strand breaks in the DNA of cells treated with angelicin were observed in normal cells and in *rad51* mutants, but they occurred at much lower frequencies in *rad 3* mutants.[60] These single-strand breaks are apparently more stable than those observed in cells removing pyrimidine dimers, which can usually be detected only when ligase activity is deficient, e.g., at elevated temperature in temperature-sensitive ligase mutants.

For 3-CPs, the *RAD3* and *RAD6* groups showed synergism, a situation similar to that for UV damage.[62] A thorough study of the sensitivity of *RAD52* group mutants[63] showed that four of them were not particularly sensitive to 3-CPs and four were. One of the sensitive ones was *rad51*; it is not clear whether this is in conflict with the results mentioned above for angelicin, or if it reflects higher adduct frequencies used with 3-CPs.

Direct measurements using radiolabeled monoadduct formers 3-CPs and 7-methylpyridopsoralen (MPP) showed that growing *RAD*[+] cells removed their monoadducts almost completely within 4 hr, but that little or no removal of MPP adducts occurred in two different deletion mutants in the *RAD3* group.[64]

Indirect assay of the repair of monoadducts of 8-MOP was obtained by measuring the amount of increased killing brought about by conversion of monoadducts to cross-links with a second dose of UVA after removal of unbound drug.[65] The increased lethality was abolished when cells were incubated (even in buffer alone) prior to the second dose, with a half-time of about 15 min. This was observed in *RAD*[+] cells and in mutants of the *RAD52* and *RAD6* groups but not in *rad3* mutants, suggesting that only excision repair was involved in this rapid removal of the relatively small number of cross-linkable monoadducts formed. A parallel result was also obtained for increased mutagenesis brought about by this irradiation scheme.[66]

The *RAD6* mutagenic pathway also appears to play a role in processing monoadducts. A member of this group, *rad9*, is sensitive to 3-CPs and exhibits synergism with a *RAD3* group mutant.[62]

B. Repair of Cross-Links

RAD3 group mutants are also sensitive to TMP and 8-MOP[60,61,65,67-69] as would be expected merely from the fact that these compounds form monoadducts, but biochemical evidence indicates that the *RAD3* pathway also operates on cross-links. Direct assays showed that DNA cross-linking disappears upon incubation of *RAD+* cells, but not with *RAD3* group mutants.[65,68,70] After treatment with TMP, double-strand breaks were observed (using neutral sucrose gradients) in *RAD+* cells, but they were at much lower frequencies in *rad3* mutants.[60] These double-strand breaks were also observed in 8-MOP-treated cells[70] but not in angelicin-treated cells.[60] Incision in TMP-treated cells as monitored by alkaline sucrose gradients was shown to be deficient in *rad1, 2, 3, 4,* and *10* and *mms19* mutants (all of which are deficient in incision at pyrimidine dimers).[68] Miller and co-workers calculated the molecular weight of the incised DNA to correspond to roughly twice the inter-cross-link distance.[68,71] While this result suggested incision on each side of a crosslink in *E. coli* (in which only single-strand breaks were observed), its interpretation is less clear in yeast. A double-strand break at cross-links should make single strands with an average length equal to the inter-cross-link distance. Under ligase-deficient conditions the number of incisions in *RAD+* cells increased about threefold.[71] Such additional nicking, interpreted as incision at the monoadducts, was not observed in *rad1, 2,* and *4* mutants, but nicks were observed in *rad3-2* mutants. It was concluded that *rad3-2* mutants could incise monoadducts but not cross-links.[71] The apparent conflict with the observation that *rad3-2* mutants are sensitive to angelicin and are deficient in incision at angelicin adducts has not been resolved directly. The role of the *RAD3* gene product is complex however, because deletions of the cloned *RAD3* gene are recessive lethals.[72]

Mutants of the excision pathway that are only partially defective in removal of pyrimidine dimers also exhibit partial defects in repair of cross-links. A mutant (*rad14*) which incises at pyrimidine dimers but is defective at some later step removes cross-linking less efficiently than *RAD+* cells.[68] *Rad7* and *rad23* mutants exhibit some removal of both pyrimidine dimers and DNA cross-linking.[73] It was suggested that the defect in these latter mutants involves the regulation of the accessibility of damaged DNA to repair systems.

The breaks observed both under alkaline and neutral conditions in *RAD+* cells containing cross-links disappear upon incubation of the cells, but not in *rad51* mutants or in the *pso2* mutant (described below).[60,70] Except for the appearance of double-strand breaks as intermediates in cross-link repair, these results suggest a similarity to the basic mechanism proposed for *E. coli*. It appears that two repair pathways, usually regarded as separate, participate in the repair of cross-links: the incision system acts on cross-links to produce substrates that are subject to processing by a recombinational system. This first step is rapid and produces breaks that are relatively long-lived.

Miller et al. reported that haploid cells in exponential phase removed cross-linking with a half-time 30 min.[68] Magana-Schwencke et al. calculated that for *RAD+* cells the mean lethal dose was about 120 cross-links per cell, but only about 1 cross-link per cell for the *pso2* mutant.[70] Under conditions giving 10% survival, growing *RAD+* cells removed most or all of the cross-linking within 2 hr. Direct measurements have shown that about 80% of induced 8-MOP adducts are removed from growing *RAD+* within 2 hr, which suggests that considerable removal of the adducts from the DNA may accompany the initial event(s) that releases the cross-linking.[64] The extreme sensitivity of the *pso2* mutant appears difficult to reconcile with the suggestion from genetic data that the *PSO2* gene product is involved in a pathway independent of the *RAD52* pathway (discussed below) and the observation that rejoining of double-strand breaks in cross-link repair requires function of the *RAD52* pathway. Interestingly, in stationary phase cells most of the cross-links remained even after 6 hr.[70] Such persistence of DNA cross-linking is puzzling, because incision is thought to be proficient in stationary phase. Perhaps the entire process is initiated only after some interaction with a second copy of the chromosome. However, the need for two copies of the compromised information for cross-link repair is by no means proven for yeast. Averbeck et al.[74] treated cells in stationary phase with psoralen or angelicin and determined survival when plated immediately or after 48 hr of holding. Diploid cells were more resistant than haploid cells by about the same factor of 2 for both psoralen and angelicin. Further, about the same increase in survival was seen upon holding regardless of agent or ploidy. These results are not consistent with a model in which two copies of the chromosome are necessary for repair, unless the actual number of cross-links formed in these experiments was so small that lethality was primarily due to monoadducts. This seems unlikely given the difference reported for sensitivity to psoralen over angelicin (about a factor of 20 in UVA dose). Clearly the question should be reexamined using measurements of actual adduct frequency to determine the extent of removal.

The *RAD6* mutagenic pathway is also involved in processing cross-links. The *rad9* mutants are sensitive 8-MOP,[67,74] but were reported to display additivity in sensitivity with *rad3* mutants.[67]

Further genetic analysis has utilized mutants isolated specifically as 8-MOP-sensitive, designated *pso1, 2,* and *3*.[75-77] These mutants complement each other and 18 different RAD mutants. The *pso1* mutant is also sensitive to 3-CPs, UV, and ionizing radiation, but does not exhibit induction of mutation by UV or furocoumarins and is epistatic to the *RAD6* group for all these agents. It thus may be allelic to *rev3-1*, a member of that group.[78] With respect to 8-MOP, it is epistatic to both the *RAD52* group and *pso2* but, unlike *rad9*, it displays synergism with *rad3*. The *pso2* mutant is also sensitive to bifunctional nitrogen mustards, but not to 3-CPs, MPP, UV, or ionizing radiation. It exhibits a slight deficiency in rate of removal of MPP adducts. Induced mutation is reduced in *pso2* for furocoumarins and nitrogen mustard but not for UV. The *pso3* mutant is sensitive to furocoumarins but not to any of the other agents mentioned, and it exhibits a more complex pattern of enhanced mutagenesis.

Taking into consideration the survival characteristics of many double mutant combinations, Henriques and Moustacchi[79] formulated the following scheme. The excision pathway and a pathway using at least the *PSO1* gene product can independently act on adducts to produce a substrate that can in turn be processed by any one of three different pathways: the *RAD6* pathway, the *RAD52* pathway, or a pathway involving the *PSO2* gene product. This is consistent with the biochemical evidence that the double-strand breaks observed during cross-link repair are rejoined poorly in *RAD52* group mutants[68] and *pso2*.[70] For pyrimidine dimers, the *PSO1* product is involved only in the *RAD6* pathway, and is needed subsequent to the action of the *RAD6* gene product. A more recent publication[78] describing interactions between *pso* and *rad* mutants for 3-CPs strengthens the idea that the *PSO1* gene product participates in the same pathway for cross-link repair as the *PSO2* or *RAD52* pathways, but is in a separate pathway for monoadduct repair. It was also concluded that the *pso2* mutant is deficient in induced recombination, rather than being generally deficient in recombination. Although these inferences are less satisfying than biochemical evidence, they do suggest the complexity of systems that process furocoumarin adducts.

As described above, the removal of potentially lethal damage (cross-linkable monoadducts) that can be inferred from changes in lethal effects of a second UVA dose is dependent only on the excision system. However, in excision-deficient *rad2* and *rad3* mutants, some reduction of the killing by the second dose was obtained if cells were allowed to proceed through one round of replication.[65] The reduction was roughly that expected if bypass of lesions, followed by segregation of the daughter DNA strands, effectively reduced the lesion frequency in the daughter cells. This phenomenon was observed in cells additionally carrying mutations in genes of either the *RAD52* or the *RAD6* pathways.[69] Thus either of these pathways can act to allow the imputed bypass of lesions. This is consistent with the notion that either recombinational gap-filling processes or some special DNA synthetic pathway for replication on damage-containing templates can overcome the block to normal replication posed by bulky adducts.

A major difficulty in using survival data to make inferences about cross-link repair is the unavoidable presence of monoadducts. One way to lessen the contribution of monoadducts is to use the reirradiation protocol to maximize cross-links. This also allows survival studies to be conducted at rather low overall adduct frequencies, because of the increased lethality of cross-links compared to monoadducts. Chanet et al. presented a large amount of survival data obtained under such conditions.[65] An important conclusion from this study is that important other effects of the necessarily large second UVA dose, such as induction of new lesions and interference with repair itself, can complicate the interpretation. Besides implicating all three epistasis groups in repair of cross-links, this study also showed that in *rad3-12*, the second UVA dose had no additional killing effect, a result obtained with other *RAD3* group mutants (*rad3-e5* and *rad2-6*) only when combined with a mutation in each of the other two epistasis groups. The triple mutant was also as sensitive to 3-CPs as to 8-MOP under conditions which maximize the cross-link to monoadduct ratio, suggesting that in these mutants monoadducts are as lethal as cross-links.

The double-irradiation protocol has also been used to study mutagenesis in repair-deficient cells.[80] In excision-deficient cells or *rad52* mutants, the second irradiation led to an increase in forward mutation that was dependent on the amount of the initial dose of radiation. Neither *rad6* nor *pso2* mutants showed the increase. The conclusion drawn from these data was that a minor pathway, independent of the excision pathway, can promote cross-link repair in a way that is mutagenic, and that both the *RAD6* and *PSO2* gene products are needed. As is the case with *E. coli*, one can only speculate upon the mechanism of incision in such a pathway. Recently, a small amount of removal of MPP adducts from both *rad1* deletion mutants and *rad2 rad6 rad52* triple mutants was reported as additional evidence for a minor excision pathway.[64] Further experiments would be required to determine whether the loss is the result of an active process.

An additional complication in drawing conclusions from studies of repair-deficient mutants arises from the need to estimate how a given pathway might act on lesions normally repaired by some other pathway. This is especially true when one attempts to interpret aspects of mutation induction, which are different for *E. coli* and yeast. This topic is addressed in a recent review.[81]

The wealth of genetic data available with yeast is confusing, and its interpretation in terms of pathways is difficult and based upon some simple assumptions which may not be entirely correct. The evidence does suggest overall similarity to the processing of monoadducts and cross-links by *E. coli*, and further study of the *pso2* and *pso3* mutants may help in understanding the pathways involved specifically with cross-links. The biochemical analysis reported so far makes a compelling case for the involvement of double-strand breaks in repair of cross-links, but whether these breaks are formed directly by the incision system or appear during further processing of single-strand breaks is unknown. Biochemical analysis needs to be extended to some of the newer mutants. There appears to be no solid biochemical evidence for participation of two copies of a sequence in cross-link repair; this issue should be addressed more directly than it has been in the past, perhaps making use of plasmids containing adducts at specific sites.

VI. REPAIR IN MAMMALIAN CELLS

Almost all of the available data on repair in these multicellular organisms are derived from experiments with explanted cells that will grow in culture. Unlike bacteria and yeast, the somatic cells from which cultures are derived do not play a direct role in reproduction of the organism. They represent but a single cell type, selected in many cases by the investigator for properties, like rapid growth, that make them resemble single-celled organisms. The factors that regulate their function *in situ* cannot be duplicated exactly in culture. Thus, attributing repair properties of cultured cells to those of the whole organism must be made with caution. Progress is being made in elucidating repair processes in *Drosophila*, an organism suitable for sophisticated genetic analysis. However, most investigations of repair have involved rodent or primate cells, and genetic analysis has been limited to the selection of repair-deficient rodent cells in culture or the use of cells from humans with diseases that may involve repair deficiency. Most of the available mutants are defective in some aspect of excision repair. Many studies of the rate and extent of removal of furocoumarin adducts and of the DNA synthesis step of excision repair have been reported for cultured mammalian cells, in contrast to work with bacteria and yeast. Attention has also been paid to the manner in which different chromatin forms influence formation and repair of furocoumarin DNA adducts.

A. Studies with Repair-Proficient Cells
1. Removal of Adducts and DNA Cross-Linking

The reported rate and extent of the removal of labeled furocoumarin adducts by different cell types vary considerably (Table 1). In all these studies the measurements reflect predominantly removal of monoadducts; cross-links, where present, were at low frequencies. Rodent cells appear to remove eventually nearly all the adducts, and the rate of removal of 8-MOP adducts in Chinese hamster ovary (CHO) cells seemed significantly greater than that for the other compounds in guinea pig cells. With primate cells, most investigators have observed a maximum removal of about 50% of the adducts. In one report,[87] extensive and very rapid removal of angelicin and methylangelicin adducts was observed. This does not appear to be a general property of repair of these particular adducts, as judged by several studies of repair replication (see below). One report[89] suggested that at high enough levels of TMP adducts, no removal occurs.

Interestingly, the observations for extent of removal of these adducts contrast with those obtained for pyrimidine dimers. In culture, rodent cells generally remove only a minor fraction of pyrimidine dimers from their genomes, while human cells remove 80 to 100% within 48 hr. How these differences may relate to intragenomic repair heterogeneity is discussed later. However, it is possible that they actually reflect the degree to which the DNA was purified from other cell components that bind furocoumarins. Most of the references for rodent cells give insufficient detail to assess the degree of purification of the DNA. In at least some of the studies with primate cells, DNA was purified through CsCl density gradients, and our own data for HMT adducts were obtained using HPLC.

With HPLC we determined that green monkey cells removed all of the various HMT monoadduct species to about the same extent.[86] In these experiments the cross-link frequency was too low to allow accurate estimation of their removal. For human cells, we have used a reirradiation protocol to increase the cross-link frequency and observed about the same extent of removal of cross-links as for monoadducts in 48 hr.[134] More extensive examination of the removal of the various species of adducts from different cell types is clearly needed.

The kinetics of removal of DNA cross-linking have been measured using a variety of techniques. Chandra et al.[83] measured removal of cross-linking in guinea pig skin treated in vivo with psoralen by determining the fraction of DNA that renatured rapidly from optical measurements. They reported a decline in the amount of cross-linking of about 25% in 72 hr. Given the lack of knowledge of the molecular weights of the DNA in the various samples and the insensitivity of the method, the results may not be very quantitative. Using a method that detects cross-links between one DNA strand and a complementary one labeled with BrdUrd, Grunert and Cleaver[90] measured removal of 8-MOP cross-linking in growing SV40-

Table 1
REMOVAL OF FUROCOUMARIN ADDUCTS BY MAMMALIAN CELLS

Cell type[a]	Damaging agent	Initial adducts per 100 kb DNA[b]	Time (hr)	% Removed	Ref.
Guinea pig skin	TMP	1.5	12	50	82
			24	75	
			48	66	
			72	92	
	Psoralen	1.1	24	55	83
			48	70	
			72	75	
Guinea pig fibroblasts	8-MOP	—[c]	24	90	84
Chinese hamster	TMP	1.2(11%)	5	50	85
			8	80	
Green monkey	AMT[d]	0.8	24	30	86
	HMT[e]	0.6(3%)	8	20	
			24	30	
			48	40	
Human fibroblasts	Angelicin	0.3	4	50	87
			24	60	
	Methylangelecin	—	4	70	
			24	80	
	7-MPP	8.3	4	10	88
			24	20	
			48	40	
	8-MOP	1.5	4	12	
			24	22	
			48	45	
	HMT	0.3(15%)	48	40	134
Human breast carcinoma	TMP	5.4(30%)	24	50	89
		>10	24	0	

[a]   Except for the first entry, experiments were with cultured cells.
[b]   Total adducts. If reported, the percent of adducts as cross-links is indicated in parenthesis.
[c]   Conversion of values reported as counts per minute per cell to adducts per base pair results in a value too high to be correct.
[d]   Aminomethyltrimethylpsoralen.
[e]   Hydroxymethyltrimethylpsoralen.

transformed human fibroblasts. Cells with an initial level of about 0.6 cross-links per 100 kilobasepairs (kbp) of DNA removed about 70% in 15 hr and 90% by 25 hr. Using a single-strand-specific nuclease to quantitate rapidly renaturing DNA, we observed 40 to 50% removal of 8-MOP cross-linking in normal human fibroblasts in 12 hr, but little additional removal by 24 hr.[91] In our study the cross-link frequencies were probably about two to three times greater than those in the study by Grunert and Cleaver. Prager et al.[89] measured cross-link removal in TMP-treated human breast carcinoma cells in stationary phase, also using a nuclease technique. They found about 50% removal in 24 hr when the initial frequency was about 3.6 per 100 kbp, but no removal at twice this frequency. Poll et al.[92] used denaturation and renaturation of DNA in gently lysed human fibroblast strains, followed by sonication and analysis by chromatography on hydroxylapatite. They reported considerable variation in results among a number of experiments, but concluded that 24 to 48 hr was required for removal of about half the cross-linking. There is one report of very rapid removal of cross-links,[93] assayed by changes in sedimentation behavior of purified DNA. However, the cross-link frequencies appeared to be much higher than in most of the other studies, and the possibility of nonspecific nicking of the DNA in these experiments must be considered.

Indirect measurements of removal of cross-linkable 8-MOP monoadducts have been made in mouse lymphoma cells, using reirradiation to detect the effect of converting monoadducts to cross-links.[94] The effect of the second irradiation on induced sister chromatid exchanges and formation of micronuclei was monitored, because it had been shown[95] that cross-links are *ten times* more effective than monoadducts in inducing such events in these cells. This approach indicated that about 60% of the cross-linkable monoadducts were removed in 6 hr, and removal appeared complete by about 24 hr. However, even at 48 hr, the second irradiation increased the number of exchanges and micronuclei, indicating that some monoadducts remained. No attempt was made to estimate the initial adduct levels, but they must have been very low, judging from the recovery of cell division. In two other studies using sublethal doses of 8-MOP or TMP, recovery of DNA synthesis and cell growth took several days, both for rodent[84] and human[96] fibroblasts. All of these studies suggest that even at low doses, biological effects of furocoumarin adducts persist long after active adduct removal can be measured.

2. Properties of DNA in Cells Undergoing Repair

The presence of monoadducts does not appear to result in easily measurable single-strand breaks, suggesting that, as with repair of UV damage, the period required for repair synthesis and ligation after the incision event is relatively short. However, in the presence of alpha polymerase inhibitors such as aphidicolin and cytosine arabinoside, breaks were detected in cells treated with methylangelicin, although the maximal levels obtained were lower than can be observed in UV-irradiated cells.[97] Using alkaline elution, Bredberg et al.[98] did not detect breaks in stationary phase human fibroblasts after a single low dose of 8-MOP, but did detect them in cells that had been reirradiated to convert monoadducts to cross-links. The frequency of these breaks increased with incubation time for up to 24 hr and were still apparent after 7 days. Using a less sensitive assay (sedimentation in alkaline sucrose gradients) to detect strand breaks, Carter et al.[99] observed that the rapid sedimentation characteristic of cross-linked DNA[91] had disappeared after a 24-hr incubation of TMP-treated fibroblasts irradiated with a single UVA dose. A problem in interpreting the results of such studies is that the initial frequencies of the different types of adducts are unknown.

3. Repair Synthesis

Measurements of the amount of repair synthesis in repair-proficient primate cells after treatment with furocoumarins have been used for comparisons with the repair following UV irradiation. Again, the predominant activity reflected by these measurements is repair of monoadducts. To determine whether the amount of repair synthesis reflects the number of lesions removed, we measured the extent of synthesis at the repaired sites for 8-MOP,[91] angelicin,[91] and 4'-aminomethyl-4,5',8-trimethypsoralen (AMT),[100] and found patch sizes of about 35 nucleotides, the same as that for UV. The time course for repair synthesis elicited by these agents in both human[101] and African green monkey cells[100] was not significantly different; repair synthesis was 50% complete by about 10 hr and had ceased by 48 hr. These measurements were done with continuous labeling, with the first point at 4 hr after treatment. Using treatment conditions similar to ours, Grunert and Cleaver[90] measured repair replication in angelicin-treated cells in four successive 2-hr intervals, terminating the measurements at 8 hr. They noted that the rate in the first 2 hr was about twice that found in the succeeding intervals, in which the rate did not decline significantly. Although these observations may appear to be in conflict, if one considers 4-hr intervals, Grunert and Cleaver found repair in the first 4 hr to be about 60% of that in the first 8 hr; this is not inconsistent with our measurements. We have not studied repair rates at very early times after furocoumarin treatment. Measurements of UDS have also demonstrated repair activity induced by furocoumarins in normal human lymphocytes[102] and rat lens epithelial cells.[103]

At saturating drug concentrations and moderately high UVA doses the initial rate of repair synthesis obtained with both 8-MOP and angelicin was only about 20% of the maximum obtained with UV.[90,91] A similar difference also was observed for both these compounds using UDS[104] and can be inferred for methylangelicin from measurements of single-strand break frequencies in cells incubated during repair with alpha polymerase inhibitors.[97] By irradiating cells in the cold to prevent repair from taking place during the irradiation, we were able to use long irradiation times to increase the number of adducts formed with furocoumarins that are only slightly soluble. With angelicin, the initial repair rate could be increased to about half the rate for UV, but with 8-MOP it could not be increased beyond about 20% of the maximum rate for UV.[101] The maximum repair rate for AMT, a much more soluble derivative with high binding efficiency, was only about 30% of that for UV. In this case, we estimated adduct frequencies indirectly by measuring cross-link frequency in purified DNA before and after reirradiation with high doses of UVA. The maximum repair rate was attained at a high monoadduct to cross-link ratio (>80:1) and at adduct frequencies well below saturation.[101] This suggests that the lower maximum repair rate is not related to formation of cross-links. For both UV and AMT the repair rate was found to reach its half-maximum at about the same adduct frequency, although the two rates themselves are different. This might indicate that the dose response for some preincision process is the same for furocoumarin adducts and pyrimidine dimers, while the kinetics for the actual incision process are different.

When both UV lesions and furocoumarin adducts are introduced at levels that result in maximum repair rates when each is used alone, one might conclude that the pathways for repair of the different adducts are separate if the repair rates appear additive. Such experiments combining UV with angelicin, AMT, or 8-MOP[101,105] showed a slight decrease in the initial rate of repair, consistent with the idea that the pathways for the removal of UV and furocoumarin monoadducts are not independent.

Because one cannot form cross-links exclusive of monoadducts, a direct examination of repair synthesis that might accompany removal of cross-links is precluded. We[101] and Gruenert and Cleaver[90] studied the effect on repair synthesis of conversion of 8-MOP monoadducts to cross-links by the reirradiation protocol, choosing a first dose that results in repair rates significantly lower than the maximum obtainable. We detected no significant effect on repair replication of a second dose that increased cross-linking by about threefold, whereas they noted a large decrease in repair synthesis. The decrease was too large to be ascribed merely to removing monoadducts from the repairable population, and they concluded that cross-links themselves can inhibit repair. However, in our experiments with combined UV and furocoumarin treatments we found that reirradiation with UVA after washing out the unbound AMT or 8-MOP reduced the repair only slightly, as did greatly increasing a single UVA dose. The origin of the discrepancy is unknown, and the mechanism by which cross-links might inhibit repair of other furocoumarin adducts, but not pyrimidine dimers, is unclear. Studies of removal of adducts themselves at different monoadduct to cross-link ratios would be necessary to resolve these issues.

4. Recombination in Cross-Link Repair

We have no biochemical evidence for any of the steps in cross-link repair beyond the disappearance of DNA cross-linking itself. As we shall see in the next section, we know of no available mammalian cell mutants with well-characterized defects in recombinational processes to give clues as to the nature of the postincision events. The observation that CHO cells are most resistant to treatment with TMP in the latter part of S phase[106] could result from facilitation of cross-link repair by the presence of two daughter copies of the early replicated, active DNA. The efficient induction of sister chromatid exchanges by cross-links and their increased induction in cells thought to be defective in some aspect of cross-link repair (see below) suggests that recombination between sister chromatids may be an important step in the process. The examination of mammalian cells with only one copy of chromosomal information would seem to be precluded, but studies with viruses can be carried out at different multiplicities. For adenovirus-2[107] and SV40,[108] it seems clear that in cells infected at low multiplicity a single cross-link in the viral genome is lethal. For herpes simplex virus, conflicting reports have appeared. One study suggested monoadducts but not cross-links were repaired,[104] but others suggest that repair functions are specified by the virus.[109,110] Enhanced survival of damaged virus when many copies are introduced per cell has been demonstrated for herpes virus[109] and SV40.[108] The latter is perhaps more pertinent to cell-coded functions, since SV40 does not itself specify recombination or repair functions. These results are consistent with a requirement for two copies of a sequence for cross-link repair.

5. Tolerance Mechanisms

Although the mechanisms whereby mammalian cells overcome blocks to replication posed by lesions are not understood in any detail, the phenomenon has been demonstrated.[13,18] We are aware of a single study of properties of newly synthesized DNA (other than its amount) in cells containing furocoumarin adducts. Ben-Hur and Elkind[111] found that the size of newly synthesized DNA in CHO cells treated with 8-MOP was related inversely (and linearly) to the amount of psoralen bound to the parental DNA, under conditions where few cross-links were produced. This result is similar to that obtained with UV. The size of DNA labeled during incubation with radioactive DNA precursor for 1 hr immediately after treatment was seen to increase when the cells were incubated in nonradioactive medium for an additional 3 hr. This behavior is usually interpreted as evidence for recombinational gap filling or translesion synthesis in gaps left opposite lesions. In our laboratory Vos has recently obtained evidence for bypass of monoadducts in human cells by demonstrating cross-linkable sites in the hybrid DNA made by cells incubated in density label after treatment with HMT.[135]

B. Studies with Mutant Cells

Removal of both monoadducts and cross-links occurs in repair-proficient cells, and many aspects of repair of DNA containing monoadducts resemble those of the removal of UV-induced damage. Further elucidation of repair mechanisms in mammalian cells will require identification of the genes involved to allow examination of their products, which in turn depends upon isolation and study of cells deficient in repair processes.

1. Rodent Cells

Most of the repair-deficient mutants of cultured cells have been obtained by screening mutagenized CHO cells for sensitivity to UV and alkylating agents.[112-114] Many mutants showing strong sensitivity to UV have been obtained and classed into five complementation groups by measuring the survival of cell hybrids after UV radiation. Cells from each of these groups are also sensitive to a number of other agents that produce bulky adducts. They fail to incise DNA after UV irradiation, perform little or no repair synthesis, and do not remove significant amounts of adducts formed by 7-bromomethylbenz(a)anthracene. Thus, they appear to resemble the *uvr* mutants of *E. coli* and the *RAD3* group mutants of yeast. However, they differ in their degree of sensitivity to a variety of cross-linking agents. Although all the groups are more sensitive than the repair-proficient cells to cross-linking agents, groups 2 and 4 exhibit extreme hypersensitivity to alkylating cross-linkers. Cells of group 2 have been shown to be much more sensitive to 8-MOP than repair-proficient or group 1 cells.[115]

A mutant, EM9, that is hypersensitive to methyl and ethyl donors, and slightly sensitive to X-rays is no more sensitive to 8-MOP (or UV) than repair-proficient cells. Although there are mutants that appear to be deficient in repair of double-strand breaks, as far as is known they have not yet been tested for sensitivity to furocoumarins.

A variety of mouse lymphoma cells that are sensitive to DNA-damaging agents have also been isolated. Like the CHO cells there appear to be several complementation groups of mutants deficient in the incision step of bulky adduct repair, and mutants sensitive to other agents have been characterized. A good summary of the derivation and properties of these mutants is found in Reference 114. As yet, no reports concerning the processing of furocoumarin adducts by these cells have appeared.

2. Human Cells

Patients with hereditary (and usually cancer-prone) diseases have been the source of cultured cells that display sensitivity to DNA-damaging agents.[10,13] The disease xeroderma pigmentosum (XP) has been a rich source of such mutant cells. With the exception of one group (the XP variants), the cells display varying degrees of deficiency in early steps of excision repair of bulky adducts and have been classified into nine complementation groups based upon levels of UV-induced UDS in heterokaryons. Cells of complementation group A typically show almost a total lack of incision after UV irradiation and do not remove pyrimidine dimers or a number of other bulky adducts from their DNA. Many studies have indicated that these cells are deficient in repair of furocoumarin adducts. Angelicin and methylangelicin adducts are poorly removed,[87] and repair synthesis in these cells is almost totally lacking after treatment with 8-MOP[82,90,91] or angelicin.[90,91] Removal of DNA cross-linking is severely deficient,[90] and the stable breaks associated with repair of cross-links detected by alkaline elution[98] were not observed in XP-A cells. These cells (as well as group D cells) also appear unable to perform host cell reactivation of adenovirus containing TMP adducts[107] or herpes virus containing angelicin monoadducts.[104,110] A straightforward interpretation of all these results is that a single endonuclease activity is required for incision at both monoadducts and cross-links, as seems to be the case for *E. coli*. However, it is entirely possible that the defective activity in these cells involves identification of adducts or rendering them accessible to further processing, as extracts from these cells can promote pyrimidine dimer excision from pure DNA.[13]

Processing of furocoumarin adducts in the other complementation groups of XP has not been studied as intensively. As judged by formation of breaks in the presence of polymerase inhibitors, cells of complementation groups C and D are also deficient in repair of methylangelicin adducts, and the relative deficiency is about the same as for UV damage.[97] Removal of angelicin adducts appeared to be much more proficient than removal of pyrimidine dimers in cells of group D and slightly more proficient in cells of group E. Interestingly, our measurements of cross-linking in an XP variant strain[91] appeared to show a slight deficiency compared to a normal strain. This is an area that deserves further study. No comparisons have been reported of sensitivity of the different XP complementation groups to UV and cross-linking furocoumarins, to determine if any resemble the CHO group 2 or 4 cells. This appears unlikely, however, because cells of these groups are complemented by cells of all of the XP groups so far tested.

Carter et al.[99] observed persistence of rapidly sedimenting DNA after TMP treatment of fibroblasts cultured from two unrelated patients with the rare disease dyskeratosis congenita (DKC). Leukocytes from these patients exhibited higher levels of sister chromatid exchanges after TMP treatment than normal cells. However, using cells derived from different patients, other investigators were unable to confirm increased sensitivity of DKC cells to cross-linking agents or increased sister chromatid exchanges in them.[93] The rate of removal of cross-linking was, however, slightly slower in these DKC strains.

Cells from patients with Fanconi's anemia (FA), which shares some characteristics with DKC, have been studied much more extensively. Many of the FA strains are hypersensitive to mitomycin C (MMC), a commonly used alkylating agent which makes adducts to guanine and can also form cross-links between guanines in complementary strands. The initial report suggesting that these cells were generally defective in cross-link removal was followed by a number of contradictory reports concerning removal of both MMC and psoralen cross-links. For example, Poll et al.[92] showed that two different FA strains, both of which were very sensitive to MMC, were only slightly sensitive to TMP, and neither could be demonstrated to exhibit a consistent lack of removal of MMC or 8-MOP DNA cross-linking. A recent report by Moustacchi and Diatloff-Zita[116] contains an excellent summary of the contradictory results concerning the defect in these cells and provides new insights into their possible cause. They found that FA cells differ in their capacity to recover normal rates of DNA synthesis after moderate doses of 8-MOP plus UVA. One class (six different cell strains) appeared normal, while the other (four strains) did not recover. It was suggested that the first class is able to remove cross-links, but the second is defective at some step, possibly after the incision, that prevents resumption of DNA synthesis. Interestingly, the strain CRL1196 was in the former class; a source of some of the controversy is the evidently normal behavior of this strain, e.g., our own demonstration of its removal of 8-MOP cross-linking[91] compared to the lack of removal of cross-linking observed in an SV40-transformed FA line.[90] A recent report does indicate that the increased sister chromatid exchange attributed to psoralen cross-links is shared by both classes.[117] It appears that we must await classification of the many different strains used in the various studies to determine how well these two classes correlate with accounts of defective repair and increased sensitivity to cross-linking agents. These classes may correspond to the two complementation groups of FA recently identified by examining cytotoxicity and chromosome breakage induced in hybrids by MMC.[118] It seems likely that FA cells will eventually provide a system for study of both incision and postincision events in cross-link processing and may help elucidate differences in processing of different kinds of cross-links.

C. Repair in Specific DNA Sequences and in Relation to Chromatin Structure

In all of the foregoing discussion, the DNA in a cell has been considered as a uniform substrate both for the production of damage and its repair. However, it has long been recognized that eukaryotic chromatin partitions DNA into different structural and functional types, probably serving in large part to regulate accessibility to cellular machinery. How chromatin structure influences the organization and function of repair systems has been of interest to many investigators.[18,119] Psoralens were among the first DNA-damaging agents to be shown to produce a nonrandom distribution of adducts in DNA, having a strong preference for binding to the nucleosome linker DNA rather than to the core DNA. This was originally demonstrated both by using electron microscopy to visualize the positions of cross-links along DNA from psoralen-treated chromatin[120] and by using staphylococcal nuclease to show that the sensitivity of labeled psoralens in chromatin corresponded to that of the linker DNA.[121] More recent digestion studies have indicated that for angelicin and methylangelicin the relative proportion of adducts in linker and core DNA is a function of the total amount of binding, with the difference most pronounced at high adduct frequencies.[122]

A series of studies by Lieberman and co-workers showed that in human cells the repair patches synthesized in repsonse to UV were initially much more sensitive to digestion by staphylococcal nuclease than the bulk of the DNA in chromatin and that the enhanced sensitivity disappeared within a few hours.[123] The initial sensitivity could result from an alteration of the chromatin to allow the repair enzymes to act on the DNA. If the nucleosomes were then restored to their original positions along the DNA, the sensitivity of repair patches might then resemble the DNA as a whole simply because their distribution, like that of the pyrimidine dimers they replaced, would be random with respect to the positioning of nucleosomes. To examine the process in more detail, we repeated these analyses using angelicin as a damaging agent, to bias the placement of damage toward the nuclease-sensitive linker DNA.[124] We determined directly that our conditions resulted in a much higher nuclease sensitivity of angelicin-thymine adducts when chromatin in nuclei was digested, but not when purified DNA was. Pyrimidine dimers did not exhibit increased sensitivity. The level of preferential localization of adducts in nuclease-sensitive chromatin is consistent with the high adduct frequencies we used to maximize radioactivity in repair patches. We found essentially the same changes in nuclease sensitivity for repair synthesis after angelicin treatment and after UV irradiation. Even though the repair patches made in response to angelicin should have been initiated in linker DNA, their nuclease sensitivity declined rapidly with time. Since the repair patches are small compared to the length of core DNA, we concluded that the decline in sensitivity could not reflect local reformation of nucleosomes at precisely their original positions. This return of repair patches to normal sensitivity was also observed by Cleaver,[122] apparently using conditions under which the adducts were more randomly distributed. He also showed that methylangelicin adducts in chromatin themselves become less sensitive to nuclease with time, even in XP-A cells, in which little or no repair takes place. However, the randomization of adducts occurred much too slowly to account for the rate of decline in nuclease sensitivity of the repair patches.

Studies of alpha DNA of African green monkey cells (reviewed in Reference 125) have provided insights about the influences of a higher level of chromatin structure on DNA repair. Alpha DNA is a nontranscribed, highly repetitive DNA found primarily near chromosome centromeres in constitutive heterochromatin. It comprises 15 to 20% of the genome of these cells, but no function for it has been identified. Highly homologous sequences are found in other primate genomes, but in much lower quantities. Alpha DNA in green monkey cells is organized as long arrays of 172-bp repeating units, most of which contain a single recognition site for the restriction endonuclease Hind III. This allows isolation of alpha DNA from cellular DNA by separation of Hind III fragments on preparative 2% agarose gels. The average array length was recently estimated to be at least 450 monomer units (about 80 kb).[126]

In stationary phase cells, repair synthesis in alpha in response to UV resembles that in the bulk of the DNA, but the repair synthesis in response to angelicin, AMT, or HMT is only 20 to 30% of that in the bulk DNA from 4 to 48 hr after treatment. Using labeled compounds, it was shown that this reflects deficient removal of the adducts themselves.[86,100] This deficiency is not the result of preponderant formation in alpha DNA of some poorly repaired adduct. HPLC analysis of HMT adducts[86] showed that the relative proportions of the various monoadducts formed, including the diastereomers of furan-T monoadducts, were the same in alpha and bulk DNA, regardless of whether pure DNA or cells were the substrates for adduct formation. Using treatment conditions that gave only about 3% of the adducts as cross-links, we also found that in 48 hr all the monoadduct species were removed from bulk DNA to about the same extent, roughly 40%. The deficiency observed with angelicin and similar results with aflatoxin B_1 eliminated cross-links as contributing to the defect and suggests that something about its chromatin structure hinders repair of certain types of damage in alpha DNA. Although we observed similar total adduct frequencies for bulk and alpha DNA, the ratio of cross-links to monoadducts formed in vivo was considerably lower in alpha than in the bulk DNA. This lower apparent quantum yield for conversion of monoadducts to cross-links in alpha in vivo could be due to a more condensed chromatin structure, restricting the flexibility of the nucleosomal linker DNA, in which most of the psoralen adducts are found. Model-building studies[128] suggest that a large kink must occur in the DNA at the site of a cross-link; DNA with reduced flexibility might therefore exhibit a lower quantum yield for cross-link formation.

Extensive studies of repair in alpha DNA using other damaging agents, growing cells, and combinations of adducts revealed the complexity of the interactions between repair systems and this DNA. They suggest that the degree to which DNA in alpha chromatin must be made accessible to repair systems varies with the type of adduct and the growth state of the cell.[125]

Still another level of chromatin structure that has an impact on repair is the organization of transcriptionally active or potentially active DNA. Since 1984 we have concentrated on developing and using methods to study repair in sequences in the cell that cannot be isolated in substantial amounts, such as active genes and their surrounding sequences.[125,129] We have been able to quantitate removal of pyrimidine dimers from sequences present in as few as two copies per cell using T4 endonuclease V to incise DNA specifically at the lesions and hybridization with specific-sequence ^{32}P-labeled probes to detect the genomic sequences of interest. With purified, restricted DNA the fraction of such molecules resistant to the endonuclease is measured using Southern hybridization analysis of DNA separated by electrophoresis on alkaline gels. This fraction allows calculation of lesion frequency using the Poisson equation. Repair in stretches of DNA spanning a particular sequence is measured by treating permeabilized cells with the endonuclease, resolving their DNA on alkaline sucrose gradients, and probing the gradient fractions for specific sequences. The molecular weight distribution of total DNA is compared to the DNA that hybridizes to different probes to examine differences in removal of pyrimidine dimers. This technique can be used only where little or no DNA replication occurs during the repair incubation, as newly synthesized lesion-free DNA cannot be distinguished from repaired DNA. With purified restriction fragments, density labeling is used to eliminate replicated DNA from the analysis.

Using these techniques we have shown that active genes in rodent cells are repaired more extensively than nontranscribed sequences or the genome as a whole, and that an active gene in human cells is repaired at a faster rate than the overall genome (summarized in Reference 125). Recent results[127] have shown this preferential repair is specific to the transcription template strand of DNA. We conclude that rodent cells in culture, which display low overall removal of pyrimidine dimers, do remove them efficiently from sequences vital to cell function, and thereby achieve UV resistance comparable to that of human cells, which efficiently remove pyrimidine dimers throughout their genome. The reported differences between the rodent and human cells for the extent of removal of furocoumarin adducts (Table 1) could indicate fundamental differences in the role of chromatin structure in the recognition of furocoumarin adducts and pyrimidine dimers. Alternatively, there might be a large bias in rodent cells toward formation of adducts in these preferentially repaired region. Differential formation of adducts in DNA of different activity for ribosomal DNA in both *Dictyostelium*[130] and *Tetrahymena*[131] cells have been attributed to the near absence of nucleosomal structure in the transcribed DNA. The limited removal of furocoumarin adducts reported for human cells might indicate preferential repair similar to that observed for pyrimidine dimers in CHO cells.

Methods for analysis of furocoumarin adducts in specific sequences have recently been developed and used in our laboratory.[136] The detection of specific sequences of DNA is also achieved by hybridization, but quantitation of adducts relies not on the specificity of a lesion-specific endonuclease, but on the rapid renaturation of cross-linked DNA.[132] For specific restriction fragments, electrophoresis on agarose gels is used to resolve cross-linked molecules, which migrate as double-stranded DNA after denaturation-renaturation from cross-link-free molecules, whose strands migrate as single-stranded DNA. Poisson analysis can then be used to calculate cross-link frequency. Cross-linkable monoadducts can be quantitated after converting them to cross-links by further irradiation in vitro. For unrestricted DNA, the single- and double-stranded DNA are well resolved by sedimentation in CsCl density gradients at a pH around 11.2. The populations of cross-linked and un-cross-linked molecules thus provided can be hybridized to various probes to determine the fraction of DNA containing cross-links in different cellular sequences. Cross-link frequencies can then be derived from the molecular weight of the DNA, determined by sedimentation analysis. Results obtained so far with human cells demonstrate both differential production and removal of adducts in different genes and nontranscribed sequences; studies with rodent cells are in their initial stages. We hope this line of investigation will contribute not only to an understanding of chromatin effects on repair systems, but also allow a new approach to the biochemical analysis of cross-link repair.

VII. CONCLUSIONS AND PERSPECTIVE

In all of the systems that have been studied, furocoumarin monoadducts are processed primarily by excision repair. Both genetic and biochemical evidence shows that incisions are made in the DNA by complex enzymatic systems that have a broad specificity for bulky adducts and distorting lesions. With a few exceptions, the subsequent steps of excision of the lesions, polymerization, and ligation have been demonstrated to be similar if not identical to those occurring during removal of pyrimidine dimers, the most studied bulky lesion. Detailed comparison of the capacity and affinity of repair systems for monoadducts and pyrimidine dimers, and the relative lethality and mutagenicity of the two different lesions have not been well studied, in large part due to the difficulty of accurate quantitation of the furocoumarin monoadducts. For mammalian cells there is considerable suggestive evidence that preincision aspects of repair can differ for the two lesions. Perhaps with the advent of good assays for adduct frequency, we will begin to see careful comparisons.

Our understanding of processing of cross-links is rudimentary. It is clear that the bulky adduct incising systems are necessary for cross-link removal, and that mutants of bacteria, yeast, and mammalian cells have also been found that are much more sensitive to cross-links than monoadducts. The hypothesis that recombinational processes are involved in reconstructing the compromised information has its best support in both biochemical and genetic data in the case of *E. coli*. For yeast, genetic data suggest that recombination and a mutagenic synthetic process can act independently at incised cross-links, and the biochemical data indicate that scissions on both strands of the cross-linked DNA are made during repair. However, attempts to demonstrate directly a requirement for two chromosomes for cross-link repair in both yeast and *E. coli* have been unsuccessful. This may be due to technical difficulties, or to other pathways that can substitute for recombination. Studies with extrachromosomal DNA, especially containing cross-links at specific sites, promise to be informative, but their interpretation may not always be straightforward. For example, a recent study by Roberts and Strike[133] documented that psoralen-treated plasmid survives better in uvr$^+$ cells when they already harbor a related plasmid. However, most of the effect could be ascribed to an increased time for removal of monoadducts before plasmid replication in those cells, brought about by the copy number control system.

It seems reasonable to expect that progress in understanding processing of pyrimidine dimers in all the biological systems discussed will apply to furocoumarin monoadducts in general, and that with reliable and easy assays for these adducts, comparative studies will be helpful in determining aspects of repair that may differ according to the specific nature of different bulky adducts. However, furocoumarins themselves appear to be the best model compounds at present for elucidating the nature of cross-link repair, keeping in mind the possibility that major differences exist between processing of furocoumarin cross-links and those of guanine-specific alkylating agents. A detailed understanding of this phenomenon will probably depend ultimately on progress in elucidating recombination mechanisms. In this regard, the best model system for higher eukaryotes may well turn out to be *Drosophila*. For mammalian cells, collection of data concerning removal of cross-links themselves, as opposed to removal of cross-linking, is needed, especially in mutant cells thought to be defective in postincision processes. Studies of processing in specific genomic sequences and in extrachromosomal DNA may also provide needed insights into the biochemistry of the process.

The omission of a discussion of mutagenesis in this chapter is intentional. Clearly, valuable insights into error-prone processing mechanisms occurring in cells can be gained from examining the nature of the mutations induced. The focus of this chapter has been on those mechanisms that reduce the frequency of those mutagenic events.

ACKNOWLEDGMENTS

The research conducted in our laboratory was supported by grants to P.C. Hanawalt from the National Institutes of Health of the U.S. Public Health Service, the American Cancer Society, and the U.S. Department of Energy. I am grateful to P. Hanawalt for his continued support and encouragement, and for discussions concerning this article, and to A.K. Ganesan for discussion and critical reading of the manuscript. We are indebted to J. Kaye for bringing an interest in furocoumarins to the laboratory.

REFERENCES

1. **Ashwood-Smith, M. J., Poulton, G. A., Barker, M., and Mildenberger, M.,** 5-Methoxypsoralen, an ingredient in several suntan preparations, has lethal, mutagenic and clastogenic properties, *Nature (London)*, 285, 407, 1980.
2. **Ivie, G. W., Holt, D. L., and Ivey, M. C.,** Natural toxicants in human foods: psoralens in raw and cooked parsnip root, *Science*, 213, 901, 1981.
3. **Ashwood-Smith, M. J., Ceska, O., and Chaudhari, S. K.,** Mechanism of photosensitivity reactions to diseased celery, *Br. Med. J.*, 290, 1249, 1985.
4. **Ben-Hur, E. and Song, P.-S.,** The photochemistry and photobiology of furocoumarins (psoralens), *Adv. Radiat. Biol.*, 11, 131, 1984.
5. **Song, P.-S. and Tapley, K. J., Jr.,** Photochemistry and photobiology of psoralens, *Photochem. Photobiol.*, 29, 1177, 1979.
6. **Rodighiero, G. and Dall'Acqua, F.,** Biochemical and medical aspects of psoralens, *Photochem. Photobiol.*, 24, 647, 1976.
7. **Scott, B. R., Pathak, M. A., and Mohn, G. R.,** Molecular and genetic basis of furocoumarin reactions, *Mutat. Res.*, 39, 29, 1976.
8. **Brendel, M. and Ruhland, A.,** Relationships between functionality and genetic toxicology of selected DNA-damaging agents, *Mutat. Res.*, 133, 51, 1984.
9. **Wilkins, R. J.,** Sequence specificities in the interactions of chemicals and radiations with DNA, *Mol. Cell. Biochem.*, 64, 111, 1984.
10. **Hanawalt, P. C. and Sarasin, A.,** Cancer-prone hereditary diseases with DNA processing abnormalities, *Trends Genet.*, 2, 124, 1986.
11. **Claverys, J. P. and Lacks, S. A.,** Heteroduplex deoxyribonucleic acid base mismatch repair in bacteria, *Microbiol. Rev.*, 50, 133, 1986.
12. **Moustacchi, E.,** DNA repair in yeast: genetic control and biological consequences, *Adv. Radiat. Res.*, 13, 1987, in press.
13. **Friedberg, E. C.,** *DNA Repair*, W. H. Freeman, New York, 1985.
14. **Walker, G. C.,** Inducible DNA repair systems, *Annu. Rev. Biochem.*, 54, 425, 1985.
15. **Walker, G. C., Marsh, L., and Dodson, L. A.,** Genetic analyses of DNA repair: inference and extrapolation, *Annu. Rev. Genet.*, 19, 103, 1985.
16. **Strauss, B. S.,** Cellular aspects of DNA repair, *Adv. Cancer Res.*, 45, 45, 1985.
17. **Lindahl, T.,** DNA repair enzymes, *Annu. Rev. Biochem.*, 51, 61, 1982.
18. **Hanawalt, P. C., Cooper, P. K., Ganesan, A. K., and Smith, C. A.,** DNA repair in bacteria and mammalian cells, *Annu. Rev. Biochem.*, 48, 783, 1979.
19. **Cooper, P. K.,** Characterization of long patch excision repair of DNA in ultraviolet-irradiated *Escherichia coli*: an inducible function under rec-lex control, *Mol. Gen. Genet.*, 185, 189, 1982.
20. **Wang, T. V. and Smith, K. C.,** Postreplicational formation and repair of DNA double-strand breaks in UV-irradiated *Escherichia coli uvrB* cells, *Mutat. Res.*, 165, 39, 1986.
21. **Szostak, J. W., Orr-Weaver, T. L., Rothstein, R. J., and Stahl, F. W.,** The double-strand-break repair model for recombination, *Cell*, 33, 25, 1983.
22. **Calvin, N. M. and Hanawalt, P. C.,** Electrophoretic separation of furocoumarin:DNA adducts, *Photochem. Photobiol.*, 40, 161, 1984.

23. **Calvin, N. M. and Hanawalt, P. C.,** Photoadducts of 8-methoxypsoralen to cytosine in DNA, *Photochem. Photobiol.,* 45, 323, 1987.

24. **Ganesan, A. K., Smith, C. A., and Van Zeeland, A. A.,** Measurement of pyrimidine dimer content of DNA in permeabilized bacterial or mammalian cells with endonuclease of bacteriophage T4, in *DNA Repair: A Laboratory Manual of Research Procedures,* Friedberg, E. C. and Hanawalt, P. C., Eds., Marcel Dekker, New York, 1980, 89.

25. **Smith, C. A., Cooper, P. K., and Hanawalt, P. C.,** Measurement of repair replication by equilibrium sedimentation, in *DNA Repair: A Laboratory Manual of Research Procedures,* Friedberg, E. C. and Hanawalt, P. C., Eds., Marcel Dekker, New York, 1981, 289.

26. **Bordin, F., Carlassare, F., Baccichetti, F., and Anselmo, L.,** DNA repair and recovery in *Escherichia coli* after psoralen and angelicin photosensitization, *Biochim. Biophys. Acta,* 447, 249, 1976.

27. **Grossweiner, L. I. and Smith, K. C.,** Sensitivity of DNA repair-deficient strains of *Escherichia coli* K-12 to various furocoumarins and near-ultraviolet radiation, *Photochem. Photobiol.,* 33, 317, 1981.

28. **Bridges, B. A., Mottershead, R. P., and Knowles, A.,** Mutation induction and killing of *E. coli* by DNA adducts and crosslinks: a photobiological study with 8-methoxypsoralen, *Chem. Biol. Interact.,* 27, 221, 1979.

29. **Cole, R. S.,** Inactivation of *Escherichia coli,* F' episomes at transfer, and bacteriophage lambda by psoralen plus 360 nm light: significance of deoxyribonucleic acid crosslinks, *J. Bacteriol.,* 107, 846, 1971.

30. **Sinden, R. R. and Cole, R. S.,** Repair of cross-linked DNA and survival of *Escherichia coli* treated with psoralen and light: effects of mutations influencing genetic recombination and DNA metabolism, *J. Bacteriol.,* 136, 538, 1978.

31. **Bridges, B. A. and Mottershead, R. P.,** Inactivation of *Escherichia coli* by near-ultraviolet light and 8-methoxypsoralen: different responses of strains B/r and K-12, *J. Bacteriol.,* 139, 454, 1979.

32. **Seeberg, E.,** Strand cleavage at psoralen adducts and pyrimidine dimers in DNA caused by interaction between semi-purified *UVR⁺* gene products from *Escherichia coli, Mutat. Res.,* 82, 11, 1981.

33. **Sancar, A., Franklin, K. A., Sancar, G., and Tang, M. S.,** Repair of psoralen and acetylaminofluorene DNA adducts by ABC excinuclease, *J. Mol. Biol.,* 184, 725, 1985.

34. **Strike, P., Wilbraham, H. O., and Seeberg, E.,** Repair of psoralen plus near ultraviolet light damage in bacteriophages T3 and T4, *Photochem. Photobiol.,* 33, 73, 1981.

35. **Drake, J. W.,** Photodynamic inactivation and mutagenesis by angelicin (isopsoralen) or thiopyronin (methylene red) in wild-type and repair deficient strains of bacteriophage T4, *J. Bacteriol.,* 162, 1311, 1985.

36. **Bridges, B. A. and Stannard, M.,** A new pathway for repair of cross-linkable 8-methoxypsoralen monoadducts in Uvr strains of *Escherichia coli, Mutat. Res.,* 92, 9, 1982.

37. **Miller, S. S. and Eisenstadt, E.,** Enhanced sensitivity of *Escherichia coli umuC* to photodynamic inactivation by angelecin (isopsoralen), *J. Bacteriol.,* 162, 1307, 1985.

38. **Cole, R. S.,** Repair of DNA containing interstrand crosslinks in *Escherichia coli:* sequential excision and recombination, *Proc. Natl. Acad. Sci. U.S.A.,* 70, 1064, 1973.

39. **Cole, R. S. and Sinden, R. R.,** Repair of cross-linked DNA in *Escherichia coli, Basic Life Sci.,* 5B, 487, 1975.

40. **Cole, R. S., Levitan, D., and Sinden, R. R.,** Removal of psoralen interstrand cross-links from DNA of *Escherichia coli:* mechanism and genetic control, *J. Mol. Biol.,* 103, 39, 1976.

41. **Yoakum, G. H. and Cole, R. S.,** Role of ATP in removal of psoralen cross-links from DNA of *Escherichia coli* permeabilized by treatment with toluene, *J. Biol. Chem.,* 252, 7023, 1977.

42. **Sinden, R. R. and Cole, R. S.,** Topography and kinetics of genetic recombination in *Escherichia coli* treated with psoralen and light, *Proc. Natl. Acad. Sci. U.S.A.,* 75, 2373, 1978.

43. **Cole, R. S., Sinden, R. R., Yoakum, G. H., and Broyles, S.,** On the mechanism for repair of crosslinked DNA in *E. coli* treated with psoralen and light, in *DNA Repair Mechanisms,* Vol. 9, Hanawalt, P. C., Friedberg, E. C., and Fox, C., Eds., Academic Press, Ner York, 1978, 287.

44. **Van Houten, B., Gamper, H., Holbrook, S. R., Hearst, J. E., and Sancar, A.,** Action mechanism of ABC excision nuclease on a DNA substrate containing a psoralen crosslink at a defined position, *Proc. Natl. Acad. Sci., U.S.A.,* 83, 8077, 1986.

45. **Van Houten, B., Gamper, H., Hearst, J. E., and Sancar, A.,** Construction of DNA substrates modified with psoralen at a unique site and study of the action mechanism of ABC exinuclease on these uniformly modified substrates, *J. Biol. Chem.,* 261, 14135, 1986.

46. **Grossman, L., Caron, P. R., and Oh, E. Y.,** The involvement of an *E. coli* multiprotein complex in the complete repair of UV-damaged DNA., *Basic Life Sci.,* 38, 287, 1986.

47. **Grover, N. B., Margalit, A., Zaritsky, A., Ben-Hur, E., and Hansen, M. T.,** Sensitivity of exponentially growing populations of *Escherichia coli* to photo-induced psoralen-DNA interstrand crosslinks, *Biophys. J.,* 33, 93, 1981.

48. **Zhen, W.-P., Jeppesen, C., and Nielsen, P. E.,** Repair in *E. coli* of a psoralen-DNA interstrand crosslink specifically introduced into T410A411 of the plasmid pUC 19, *Photochem. Photobiol.,* 44, 47, 1986.

49. **Belogurov, A. A., Zuev, A. V., and Zavil'gel'skii, G. B.**, Repair of 8-methoxypsoralen monoadducts and diadducts in bacteriophages and bacteria, *Mol. Biol. (Moscow)*, 10, 705, 1976.

50. **Ben-Hur, E., Prager, A., and Riklis, E.**, Measurement of DNA crosslinks by S1 nuclease: induction and repair in psoralen-plus-360 nm light treated *Escherichia coli*, *Photochem. Photobiol.*, 29, 921, 1979.

51. **Bridges, B. A. and von Wright, A.**, Influence of mutations at the *rep* gene on survival of *E. coli* following UV irradiation or 8-methoxypsoralen photosensitization. Evidence for a *recA⁺ rep⁺* dependent pathway for repair of crosslinks, *Mutat. Res.*, 82, 229, 1981.

52. **Bridges, B. A.**, Further characterization of repair of 8-methoxypsoralen crosslinks in UV-excision-defective *Escherichia coli*, *Mutat. Res.*, 132, 153, 1984.

53. **Cupido, M. and Bridges, B. A.**, Uvr-independent repair of 8-methoxypsoralen crosslinks in *Escherichia coli*: evidence for a recombinational process, *Mutat. Res.*, 146, 135, 1985.

54. **Cupido, M. and Bridges, B. A.**, Paradoxical behaviour of pKM101-inhibition of uvr-independent crosslink repair in *Escherichia coli* by *muc* gene products, *Mutat. Res.*, 145, 49, 1985.

55. **Cooper, P. K.**, Inducible excision repair in *Eschericia coli*, in *Chromosome Damage and Repair*, Kleppe, K. and Seeberg, E., Eds., Plenum Press, New York, 1982, 139.

56. **Ahmad, S. I. and Holland, I. B.**, Isolation and analysis of a mutant of *Escherichia coli* hyper-resistant to near-ultraviolet light plus 8-methoxypsoralen, *Mutat. Res.*, 151, 43, 1985.

57. **Hansen, M. T.**, Sensitivity of *Escherichia coli acrA* mutants to psoralen plus near-ultraviolet radiation, *Mutat. Res.*, 106, 209, 1982.

58. **Lawrence, C. W.**, Mutagenesis in *Saccharomyces cerevisiae*, *Adv. Genet.*, 21, 173, 1982.

59. **Haynes, R. H. and Kunz, B. A.**, DNA repair and mutagenesis in yeast, in *Molecular Biology of the Yeast*, Saccharomyces: *Life Cycle and Inheritance*, Cold Spring Harbor Laboratory, New York, 1981, 371.

60. **Jachymczyk, R. C., von Borstel, R. C., Mowat, M. R. A., and Hastings, P. J.**, Repair of interstrand crosslinks in the DNA of *Saccharomyces cerevisiae* requires two systems for DNA repair: the *RAD3* system and the *RAD51* system, *Mol. Gen. Genet.*, 182, 196, 1981.

61. **Grant, E. L., von Borstel, R. C., and Ashwood-Smith, M. J.**, Mutagenecity of cross-links and monoadducts of furocoumarins (psoralen and angelicin) induced by 360-nm radiation in excision-repair-defective and radiation-insensitive strains of *Saccharomyces cerevisiae*, *Environ. Mutagen.*, 1, 55, 1979.

62. **Averbeck, D., Moustacchi, E., and Bisagni, E.**, Biological effects and repair of damage photoinduced by a derivative of psoralen substituted at the 3,4 reaction site: photoreactivity of this compound and lethal effect in yeast, *Biochim. Biophys. Acta*, 518, 464, 1978.

63. **Henriques, J. A. P. and Moustacchi, E.**, Sensitivity to photoaddition of mono- and bifunctional furocoumarins of X-ray sensitive mutants of *Saccharomyces cerevisiae*, *Photochem. Photobiol.*, 31, 557, 1980.

64. **Magana-Schwencke, N. and Moustacchi, E.**, A new monofunctional pyridopsoralen: photoreactivity and repair in yeast, *Photochem. Photobiol.*, 42, 43, 1985.

65. **Chanet, R., Cassier, C., Magana-Schwencke, N., and Moustacchi, E.**, Fate of photo-induced 8-methoxypsoralen mono-adducts in yeast. Evidence for bypass of these lesions in the absence of excision repair, *Mutat. Res.*, 112, 201, 1983.

66. **Cassier, C., Chanet, R., and Moustacchi, E.**, Mutagenic and recombinogenic effects of DNA cross-links induced in yeast by 8-methoxypsoralen photoaddition, *Photochem. Photobiol.*, 39, 799, 1984.

67. **Averbeck, D. and Moustacchi, E.**, 8-Methoxypsoralen plus 365 nm light effects and repair in yeast, *Biochim. Biophys. Acta*, 395, 393, 1975.

68. **Miller, R. D., Prakash, L., and Prakash, S.**, Genetic control of excision of *Saccharomyces cerevisiae* interstrand DNA cross-links induced by psoralen plus near-UV light, *Mol. Cell. Biol.*, 2, 939, 1982.

69. **Chanet, R., Cassier, C., and Moustacchi, E.**, Genetic control of the bypass of monoadducts and of the repair of cross-links photoinduced by 8-methoxypsoralen in yeast, *Mutat. Res.*, 145, 145, 1985.

70. **Magana-Schwencke, N., Henriques, J. A., Chanet, R., and Moustacchi, E.**, The fate of 8-methoxypsoralen photoinduced crosslinks in nuclear and mitochondrial yeast DNA: comparison of wild-type and repair-deficient strains, *Proc. Natl. Acad. Sci., U.S.A.*, 79, 1722, 1982.

71. **Miller, R. D., Prakash, S., and Prakash, L.**, Different effects of *RAD* genes of *Saccharomyces cerevisiae* on incisions of interstrand crosslinks and monoadducts in DNA induced by psoralen plus near UV light treatment, *Photochem. Photobiol.*, 39, 349, 1984.

72. **Naumovski, L. and Friedberg, E. C.**, A DNA repair gene required for incision of damaged DNA is essential for viability in *Saccharomyces cerevisiae*, *Proc. Natl. Acad. Sci., U.S.A.*, 80, 4818, 1983.

73. **Miller, R. D., Prakash, L., and Prakash, S.**, Defective excision of pyrimidine dimers and interstrand DNA crosslinks in *rad7* and *rad23* mutants of *Saccharomyces cerevisiae*, *Mol. Gen. Genet.*, 188, 235, 1982.

74. **Averbeck, D., Chandra, P., and Biswas, R. K.**, Structural specificity in the lethal and mutagenic activity of furocoumarins in yeast cells, *Radiat. Environ. Biophys.*, 12, 241, 1975.

75. **Henriques, J. A. and Moustacchi, E.**, Isolation and characterization of *pso* mutants sensitive to photoaddition of psoralen derivatives in *Saccharomyces cerevisiae*, *Genetics*, 95, 273, 1980.

76. **Cassier, C., Chanet, R., Henriques, J. A., and Moustacchi, E.,** The effects of three *PSO* genes on induced mutagenesis: a novel class of mutationally defective yeast, *Genetics*, 96, 841, 1980.

77. **Cassier, C. and Moustacchi, E.,** Mutagenesis induced by mono- and bi-functional alkylating agents in yeast mutants sensitive to photo-addition of furocoumarins (pso), *Mutat. Res.*, 84, 37, 1981.

78. **Moustacchi, E., Cassier, C., Chanet, R., Magana-Schwencke, M., Saeki, T., and Henriques, J. A. P.,** Biological role of photo-induced crosslinks and monoadducts in yeast DNA: genetic control and steps involved in their repair, in *Cellular Responses to DNA Damage,* Friedberg, E. C. and Bridges, B. A., Eds., Alan R. Liss, New York, 1983, 87.

79. **Henriques, J. A. and Moustacchi, E.,** Interactions between mutations for sensitivity to psoralen pho-toaddition (*pso*) and to radiation (*rad*) in *Saccharomyces cervisiae. J. Bacteriol.*, 148, 248, 1981.

80. **Cassier, C., Chanet, R., and Moustacchi, E.,** Repair of 8-methoxypsoralen photoinduced crosslinks and mutagenesis: role of the different repair pathways in yeast, *Photochem. Photobiol.*, 41, 289, 1985.

81. **von Borstel, R. C. and Hastings, P. J.,** Situation-dependent repair of DNA damage in yeast, *Basic Life Sci.*, 34, 121, 1985.

82. **Baden, H. P., Parrington, J. M., Delhanty, J. D., and Pathak, M. A.,** DNA synthesis in normal and xeroderma pigmentosum fibroblasts following treatment with 8-methoxypsoralen and long wave ultraviolet light, *Biochim. Biophys. Acta,* 262, 247, 1972.

83. **Chandra, P., Biswas, R. K., Dall'Acqua, F., Marciani, S., Baccichetti, D., Vedaldi, D., and Ro-dighiero, C.,** Post-irradiation dark recovery of photodamage to DNA induced by furocoumarins, *Biophysik,* 9, 113, 1973.

84. **Pohl, J. and Christophers, E.,** Photoinactivation and recovery in skin fibroblasts after formation of mono- and bifunctional adducts by furocoumarins-plus-UVA, *J. Invest. Dermatol.*, 75, 306, 1980.

85. **Ben-Hur, E. and Elkind, M. M.,** DNA crosslinking in Chinese hamster cells exposed to near ultraviolet light in the presence of 4,5',8-trimethylpsoralen, *Biochim. Biophys. Acta,* 331, 181, 1973.

86. **Zolan, M. E., Smith, C. A., and Hanawalt, P. C.,** Formation and repair of furocoumarin adducts in alpha deoxyribonucleic acid and bulk deoxyribonucleic acid of monkey cells, *Biochemistry*, 23, 63, 1984.

87. **Cleaver, J. E. and Gruenert, D. C.,** Repair of psoralen adducts in human DNA: differences among xeroderma pigmentosum complementation groups, *J. Invest. Dermatol.*, 82, 311, 1984.

88. **Nocentini, N.,** DNA photobinding of 7-methylpyrido(3,4-c)psoralen and 8-methoxypsoralen. Effects on macromolecular synthesis, repair and survival in cultured human cells, *Mutat. Res.*, 161, 181, 1986.

89. **Prager, A., Green, M., and Ben-Hur, E.,** Inhibition of ornithine decarboxylase induction by psoralen plus near ultraviolet light in human cells: the role of monoadducts *vs* DNA crosslinks, *Photochem. Photobiol.*, 37, 525, 1983.

90. **Gruenert, D. C. and Cleaver, J. E.,** Repair of psoralen-induced cross-links and monoadducts in normal and repair-deficient human fibroblasts, *Cancer Res.,* 45, 5399, 1985.

91. **Kaye, J., Smith, C. A., and Hanawalt, P. C.,** DNA repair in human cells containing photoadducts of 8-methoxypsoralen or angelicin, *Cancer Res.,* 40, 696, 1980.

92. **Poll, E. H., Arwert, F., Kortbeek, H. T., and Eriksson, A. W.,** Fanconi anaemia cells are not uniformly deficient in unhooking of DNA interstrand crosslinks, induced by mitomycin C or 8-methoxypsoralen plus UVA, *Hum. Genet.*, 68, 228, 1984.

93. **Kano, Y. and Fujiwara, Y.,** Dyskeratosis congenita: survival, sister-chromatid exchange and repair fol-lowing treatments with crosslinking agents, *Mutat. Res.*, 103, 327, 1982.

94. **Liu-Lee, V. W., Heddle, J. A., Arlett, C. F., and Broughton, B.,** Genetic effects of specific DNA lesions in mammalian cells, *Mutat. Res.*, 127, 139, 1984.

95. **Liu-Lee, V. W., Heddle, J. A., and Arlett, C. F.,** Repair of 8-methoxypsoralen monoadducts in mouse lymphoma cells, *Mutat. Res.*, 132, 73, 1984.

96. **Cohen, S. R., Carter, D. M., and Gala, M.,** Proliferative response patterns of human fibroblasts after photoinjury with 4,5',8-trimethylpsoralen, *J. Invest. Dermatol.*, 76, 10, 1981.

97. **Cleaver, J. E., Charles, W. C., and Kong, S. H.,** Efficiency of repair of pyrimidine dimers and psoralen monoadducts in normal and xeroderma pigmentosum human cells, *Photochem. Photobiol.*, 40, 621, 1984.

98. **Bredberg, A., Lambert, B., and Soderhall, S.,** Induction and repair of psoralen cross-links in DNA of normal human and xeroderma pigmentosum fibroblasts, *Mutat. Res.*, 93, 221, 1982.

99. **Carter, D. M., Wolff, K., and Schnedl, W.,** 8-methoxypsoralen and UVA promote sister-chromatid exchanges, *J. Invest. Dermatol.*, 67, 548, 1976.

100. **Zolan, M. E., Cortopassi, G. A., Smith, C. A., and Hanawalt, P. C.,** Deficient repair of chemical adducts in alpha DNA of monkey cells, *Cell,* 28, 613, 1982.

101. **Hanawalt, P. C., Kaye, J., Smith, C. A., and Zolan, M. E.,** Cellular responses to psoralen adducts in DNA, in *Psoralens in Cosmetics and Dermatology,* Cahn, J., Ed., Pergamon Press, New York, 1981, 133.

102. **Bohnert, E., Badilatti, B., Sidler, P., and Jung, E. G.,** DNA-repair in lymphocytes after 8-MOP + UVA and UVC irradiation, *Arch. Dermatol. Res.*, 264, 299, 1979.

103. **Jose, J. G. and Yielding, K. L.,** Photosensitive cataractogens, chlorpromazine and methoxypsoralen, cause DNA repair synthesis in lens epithelial cells, *Invest. Ophthalmol. Vis. Sci.*, 17, 687, 1978.

104. **Coppey, J., Averbeck, D., and Moreno, G.,** Herpes virus production in monkey kidney and human skin cells treated with angelicin or 8-methoxypsoralen plus 365 nm light, *Photochem. Photobiol.*, 29, 797, 1979.

105. **Gruenert, D. C. and Cleaver, J. E.,** Repair of ultraviolet damage in human cells also exposed to agents that cause strand breaks, crosslinks, monoadducts and alkylations, *Chem. Biol. Interact.*, 33, 163, 1981.

106. **Ben-Hur, E. and Elkind, M. M.,** Psoralen and near ultraviolet light inactivation of cultured Chinese hamster cells and its relation to DNA crosslinks, *Mutat. Res.*, 18, 315, 1973.

107. **Day, R. S., III, Guiffrida, A. S., and Dingman, C. W.,** Repair by human cells of adenovirus-2 damaged by psoralen plus near ultraviolet light treatment, *Mutat. Res.*, 33, 311, 1975.

108. **Hall, J. D.,** Repair of psoralen-induced crosslinks in cells multiply infected with SV40, *Mol. Gen. Genet.*, 188, 135, 1982.

109. **Hall, J. D. and Scherer, K.,** Repair of psoralen-treated DNA by genetic recombination in human cells infected with herpes simplex virus, *Cancer Res.*, 41, 5033, 1981.

110. **Fendrick, J. L. and Hallick, L. M.,** Psoralen photoinactivation of herpes simplex virus: monoadduct and cross-link repair by xeroderma pigmentosum and Fanconi's anemia cells, *J. Invest. Dermatol.*, 83, 96s, 1984.

111. **Ben-Hur, E. and Elkind, M. M.,** Post-replication repair of DNA containing psoralen addition products in Chinese hamster cells, in *Cell Biology and Tumor Immunobiology*, Vol. 349, Bucalossi, P., Veronesi, U., and Cascinelli, N., Eds., Excerpta Medica, Amsterdam, 1975, 170.

112. **Thompson. L. H. and Carrano, A. V.,** Analysis of mammalian cell mutagenesis and DNA repair using in vitro selected CHO cell mutants, in *Cellular Responses to DNA Damage*, Friedberg, E. C. and Bridges, B. A., Eds., Alan R. Liss, New York, 1983, 125.

113. **Thompson, L. H. and Hoy, C. A.,** Using repair-deficient Chinese hamster ovary cells to study mutagenesis, in *Chemical Mutagens: Principles and Methods for Their Detection*, Vol. 10, de Serres, F. J., Ed., Plenum Press, New York, 1985, 285.

114. **Thompson, L. H.,** DNA repair mutants, in *Molecular Cell Genetics*, Gottesman, M. M., Ed., John Wiley & Sons, New York, 1985, 641.

115. **Hoy, C. A., Thompson, L. H., Mooney, C. L., and Salazar, E. P.,** Defective DNA crosslink removal in Chinese hamster cell mutants hypersensitive to bifunctional alkylating agents, *Cancer Res.*, 45, 1737, 1985.

116. **Moustacchi, E. and Diatloff-Zito, C.,** DNA semi-conservative synthesis in normal and Fanconi anemia fibroblasts following treatment with 8-methoxypsoralen and near ultraviolet light or with X-rays, *Hum. Genet.*, 70, 236, 1985.

117. **Billardon, B. and Moustacchi, E.,** Comparison of the sensitivity of Fanconi's anemia and normal fibroblasts to the induction of sister-chromatid exchanges by photoaddition of mono- and bi-functional psoralens, *Mutat. Res.*, 174, 241, 1986.

118. **Duckworth-Rysiecki, G. K., Cornish, K., Clarke, C. A., and Buckwald, M.,** Identification of two complementation groups in Fanconi's anemia, *Somatic Cell Mol Genet.*, 11, 35, 1985.

119. **Bohr, V. A. and Hanawalt, P. C.,** Factors that affect the initiation of excision repair in chromatin, in *DNA Repair and Its Inhibition*, Collins, A., Johnson, R., and Downes, C., Eds., IRL Press, Oxford, 1984, 109.

120. **Hanson, C. V., Shen, C.-K. J., and Hearst, J. E.,** Cross-linking of DNA in situ as a probe for chromatin structure, *Science*, 193, 62, 1976.

121. **Cech, T. and Pardue, M. K.,** Crosslinking of DNA with trimethylpsoralen is a probe for chromatin structure, *Cell*, 11, 631, 1977.

122. **Cleaver, J. E.,** Chromatin dynamics. Fast and slow modes of nucleosome movement revealed through psoralen binding and repair, *Biochim. Biophys. Acta*, 824, 163, 1985.

123. **Lieberman, M. W., Smerdon, M. J., Tlsty, T. D., and Oleson, F. B.,** The role of chromatin structure in DNA repair in human cells damaged with chemical carcinogens and ultraviolet radiation, in *Environmental Carcinogenesis*, Emmelot, P. and Krick, E., Eds., Elsevier, Amsterdam, 1979, 345.

124. **Zolan, M. E., Smith, C. A., Calvin, N. M., and Hanawalt, P. C.,** Rearrangement of mammalian chromatin structure following excision repair, *Nature (London)*, 299, 462, 1982.

125. **Smith, C. A.,** DNA repair in specific sequences in mammalian cells, *J. Cell Sci.*, Suppl. 6, 225, 1987.

126. **Madhani, H. D., Leadon, S. A., Smith, C. A., and Hanawalt, P. C.,** Alpha DNA in African green monkey cells is organized into extremely long tandem arrays, *J. Biol. Chem.*, 261, 2314, 1986.

127. **Mellon, I., Spivak, G., and Hanawalt, P. C.,** Selective removal of transcription-blocking DNA damage from the transcribed strand of the mammalian DHFR gene, *Cell*, 51, 241, 1987.

128. **Peckler, S., Graves, B., Kanne, D., Rapoport, H., Hearst, J. E., and Kim, S.-H.,** Isolation and characterization of pyrimidine-psoralen photoadducts from DNA, *J. Mol. Biol.*, 162, 157, 1982.

129. **Bohr, V. A. and Okumoto, D. S.,** Analysis of pyrimidine dimer repair in genes, in *DNA Repair: A Laboratory Manual of Research Procedures*, Vol. 3, Friedberg, E. C. and Hanawalt, P. C., Eds., Marcel Dekker, New York, 1987, in press.

130. **Sogo, J. M., Ness, P. J., Widmer, R. M., Parish, R. W., and Koller, T.,** Psoralen-crosslinking of DNA as a probe for the structure of active nucleolar chromatin, *J. Mol. Biol.,* 178, 897, 1984.

131. **Cech, T. R. and Karrer, K. M.,** Chromatin structure of the ribosomal RNA genes of *Tetrahymena thermophyla* as analyzed by trimethylpsoralen crosslinking *in vivo, J. Mol. Biol.,* 136, 395, 1980.

132. **Vos, J.-M.H.,** Psoralen adducts in defined genes, in *DNA Repair: A Laboratory Manual of Research Procedures,* Vol. 3, Friedberg, E. C. and Hanawalt, P. C., Eds., Marcel Dekker, New York, 1987, in press.

133. **Roberts, R. J. and Strike, P.,** Repair in *E. coli* of transforming plasmid DNA damaged by psoralen plus near-ultraviolet irradiation, *Mutat. Res.,* 165, 81, 1986.

134. **Zolan, M.,** unpublished.

135. **Vos, J.-M.H. and Hanawalt, P. C.,** Processing of psoralen adducts in an active human gene: repair and replication of DNA containing monoadducts and interstrand cross-links, *Cell,* 50, 789, 1987.

Chapter 8

PSORALEN PHOTOTHERAPIES

R. M. Knobler, H. Hönigsmann, and R. L. Edelson

TABLE OF CONTENTS

I. PUVA FOR PSORIASIS AND OTHER DERMATOLOGIC CONDITIONS

A. Introduction

In 1975 a new therapeutic concept was introduced into dermatologic therapy: the combination of oral psoralen (P) and long-wave ultraviolet radiation (UVA) acronymously called PUVA.[1-3] The combined action of radiation and drug results in a therapeutic effect not achieved by UVA alone. This application of the biological action of UVA brings about reactions in the skin and in components of the circulation in the upper dermal compartment.

PUVA has its roots in ancient Egypt and India, where it has been used for thousands of years in folk medicine as a remedy for vitiligo because of its strong melanogenic activity. It was reintroduced into modern medicine by El Mofty, who treated vitiligo patients with oral 8-methoxypsoralen, (8-MOP) and sunlight.[4] The availability of high-intensity UVA radiation sources made this form of therapy practicable for large-scale treatment of skin disorders, in particular of psoriasis. Psoralens can also be used topically with subsequent exposure to UVA, but systemic administration is more practical, is easier to control, has a lower incidence of undesirable side effects, and is thus most commonly employed.

Remissions of skin disease are induced by repeated, controlled phototoxic reactions. These reactions occur only when psoralens are photoactivated by UVA. Because UVA penetrates only the superficial layers of the skin, absorption of photons is predominantly confined to this tissue compartment so that PUVA can be considered as a target-related chemotherapy. PUVA may also affect circulating blood cells, but the clinical relevance of this, if any, remains to be clarified.

PUVA-induced phototoxic reactions on the skin are clinically characterized by a delayed erythema reaction resembling sunburn which, upon overdosage, may severely damage the skin by progressing to blister formation and to superficial skin necrosis. Accurate observation of dosimetry guidelines is therefore mandatory.[5]

This chapter will detail the treatment of the most important indications, namely psoriasis, cutaneous T-cell lymphoma, vitiligo, and photodermatoses. For the treatment of other PUVA-responsive diseases, not all of which are true indications for this therapy, the reader should consult recent reviews on that subject.[6-9]

In addition, we will discuss an exciting new approach to the management of disorders caused by circulating T lymphocytes. Specifically, this treatment involves photodamaging of T cells in extracorporeally routed blood.

B. Psoralens

Psoralens are tricyclic furocoumarins present in a large number of plants, but there are also some synthetic furocoumarin derivatives. The analog most commonly used in therapy is 8-MOP (methoxsalen), which is obtained from certain plants but can also be synthesized. The synthetic compound 4,5'8-trimethylpsoralen (TMP, trisoralen) is only minimally phototoxic after oral administration because of poor absorption in the intestine and is mainly used for the treatment of vitiligo as it stimulates melanogenesis even at low skin concentrations. Other natural and synthetic psoralens are being investigated for their clinical effectiveness, yet only 5-methoxypsoralen (5-MOP, Bergapten) has proven comparably effective.[10]

Psoralens intercalate between DNA base pairs in the absence of UVA radiation to form dark complexes. Absorption of photons in the UVA (320- to 400-nm) range leads to the formation of a 3,4- or 4',5' cyclobutane adduct with pyrimidine bases of DNA (monofunctional adducts with thymine or cytosine). Linear psoralens, such as 8-MOP, TMP, and 5-MOP, can absorb a second photon, by which a bifunctional adduct (interstrand cross-link) with a 5,6 double bond of the pyrimidine base of the opposite strand is formed (for detailed information see Chapter 1). Psoralen photoadducts along the DNA double helix appear to inhibit DNA synthesis and cell division. It is generally assumed, though not actually established, that the cross-links may be responsible for the therapeutic effect of PUVA in an epidermal hyperproliferative condition such as psoriasis. However, the treatment success in some other skin diseases that are not associated with hyperproliferation can hardly be explained by this molecular effect.[9] Since psoralens also photoreact with RNA, proteins, and other cellular components, such as mitochondria (see Chapter 3), and PUVA affects a variety of different cellular and, in particular, immunological functions, different mechanisms may account for the therapeutic effect in different diseases.[11] So far only cross-linking psoralens have been found effective in psoriasis. However, pigmentation can be achieved also by noncross-linking furocoumarins, such as pyridopsoralens and angelicins.[11]

C. Treatment of Psoriasis

All forms of psoriasis respond to oral PUVA photochemotherapy, and this also includes psoriatic erythroderma as well as pustular forms. The efficiency of PUVA in clearing psoriasis and maintaining remissions has been extensively documented in the literature.[7] Since psoriasis, being a genetically determined disorder, cannot be cured, relapses eventually necessitate either maintenance treatment or resumption of complete therapy cycles. As will be detailed below, there is, thus, concern regarding potential long-term hazards, in particular carcinogenesis, which may depend on the cumulative phototoxic dose delivered to the patient during a treatment period over many years.

The general principle of photochemotherapy is to hold the dose of the drug and the interval between drug administration and UVA exposure constant and to vary the UVA irradiation according to the sensitivity of the patient. A dosage of 0.6 to 0.8 mg of 8-MOP per kilogram of body weight is administered orally 1 to 3 hr before exposure, depending on absorption characteristics of the drug brand.[9] Liquid preparations of 8-MOP are absorbed faster and yield higher and more predictable serum levels than microcrystalline forms.[12]

Broad-band UVA sources are employed for photochemotherapy. The action spectrum for treatment is between 320 and 380 nm[13] More recent information indicates that maximum efficacy occurs between 320 and 340 nm.[14] Thus, the spectral power distribution of the UV system has to be known and UVA dosimetry must be adjusted accordingly.

Initial UVA irradiations are established by either skin typing[15] or minimal phototoxic dosage (MPD) testing.[5] The MPD test allows for more accurate and also more aggressive UVA doses in the initial treatment phases. Repeated exposures are required for clearing psoriasis with irradiation increments as pigmentation develops. Standardized guidelines for dosage increments and frequency of treatments have been elaborated.[7] After clearing of skin lesions the therapy is continued according to a maintenance schedule which consists of two treatments per week for 1 month and one treatment per week for another month.

Treatment protocols vary between the different study groups. However, the rate of success in clearing psoriasis is essentially similar. The protocol employed most commonly in the U.S. uses skin typing to determine the first treatment exposure dose. Patients are treated either twice or three times a week.[15] Out of 1139 patients treated with this protocol 88.2% cleared completely, and only 3% failed to respond. Similar results have been reported in another American study of 439 patients.[16]

According to the protocol employed in the European PUVA Study (EPS), the initial UVA irradiation is the patient's individual MPD.[17] Four treatments are given per week, whereby two exposures are given on 2 consecutive days followed by a rest on day 3 after which exposures are again resumed for 2 consecutive days. In the EPS comprising 3175 patients the clearing rate was also 88.8%.[17] However, the most important difference between the two protocols concerned the cumulative phototoxic dose: the mean total UVA dose required for clearing in the EPS was 96 J/cm^2 for all skin types,[17] as compared to 191 J/cm^2 for skin type I to 296 J/cm^2 for skin type IV in the so-called 16-center study (U.S. Cooperative

Clinical Trial) referred to above.[15] Thus, although the U.S. protocol reflects a more cautious and conservative approach, the total UVA irradiation required for clearing is considerably higher than with the more aggressive EPS regimen.[17] The explanation for this difference could be that higher initial UVA irradiations and more treatments per week permit a therapeutic effect before intense melanization reduces UVA penetration.

In an Austrian study it was demonstrated that some form of maintenance treatment is essential to keep patients in remission for prolonged periods of time.[7] Many investigators in Europe have not adopted a maintenance regimen as outlined above. Follow-up observations indicate that the majority of the patients so treated remain free of disease for at least 6 to 12 months.[7]

It is important to note that the type of psoriasis influences the success rate. The study groups referred to above mostly consisted of the common acute-guttate or chronic plaque type psoriasis. However, special forms of psoriasis may respond differently.

Psoriatic erythroderma is difficult to treat because patients may flare during the initial phase of treatment. This results in prolonged clearing periods as well as high failure rates ranging from 14 to 16%.[15,17] Excellent results were noted for generalized pustular psoriasis, even during an acute exacerbation with fever and systemic illness.[17,18] However, the experience of some centers has not been as favorable, but this may be due to a different treatment regimen. Pustular eruptions of palms and soles appear rather resistant to PUVA regardless whether they are true localized pustular psoriasis, nonpsoriatic palmoplantar pustulosis, or pustular eczema.[19] Oral PUVA alone can induce a slow but definite remission in many cases, but a considerable number of patients need adjunctive therapy. The combination with topically applied 8-MOP improves the therapeutic success.[20] However, many patients do not respond satisfactorily and exhibit a high recurrence rate. More recently, in psoriatic erythroderma, pustular psoriasis of von Zumbusch, and pustular eruptions of palms and soles, combinations of PUVA with other treatment measures, particularly with oral retinoids, have been used with favorable results.

The efficacy of PUVA can be markedly improved by its combination with a variety of other therapies. Combination treatments often reduce the cumulative phototoxic dose and may thus reduce the risk of long-term side effects. Adjuvant therapy with topical corticosteroids and anthralin has been tried with some success, but this was not considered a decisive advance.[21,22]

With a combination of PUVA and UVB less than half the exposures were necessary for clearing psoriasis as compared to either modality alone. Less than half the total UVA irradiation for PUVA and only 20% of the UVB irradiation are required.[23] This combination could reduce the cumulative phototoxic dose of either treatment, resulting in less cumulative damage to the skin provided that damage due to PUVA and UVB is not additive. This has not yet been fully investigated.

The combination of PUVA with oral retinoids (etretinate and isotinoin) represents a substantial advance in improving therapeutic efficacy.[24,25] Both compounds, though of only limited value in the treatment of psoriasis if used as single agents, appear to be potent accelerators of PUVA treatment. When retinoids are administered 5 days before PUVA (1 mg/kg body weight) and are continued throughout the clearing phase, the response rate of psoriasis is accelerated, and, most important, the total phototoxic PUVA dose is reduced by more than 50% compared to conventional PUVA. In addition, poor PUVA responders and more difficult psoriasis variants, such as psoriatic erythroderma, generalized pustular psoriasis (von Zumbusch), and palmoplantar pustular disease are frequently controlled by this regimen. The high efficacy of this combination treatment, also called chemophotochemotherapy or RePUVA, has been confirmed by several large-scale studies.[26-28] RePUVA has now been adopted as routine treatment in many European institutions.

It is not known whether the synergistic action of retinoids is based on a primer or potentiator function of these drugs. Phototoxicity tests have shown that retinoids do not increase UV photosensitivity.[24]

A combination of PUVA with methotrexate has also been used with success in clearing patients with severe psoriasis, particularly in those who had had treatment failures with either PUVA or UVB phototherapy alone.[29] Although the number of PUVA exposures and the total dose of UVA was significantly reduced in these patients and the total dose of methotrexate remained well below the minimum dose reported for hepatotoxicity, the long-term use of PUVA-methotrexate combinations may be a source of concern because of possible synergistic oncogenic potential.[30]

D. Treatment of Cutaneous T-Cell Lymphoma (Mycosis Fungoides)

Cutaneous T-cell lymphoma (CTCL) is a clonal proliferative disease of T helper cells which appears to be confined to the skin in its early stages. Early lesions present as erythematous and eczematous patches and plaques which later progress to fungating tumors, hence the clinical term mycosis fungoides. After a slow progressing, chronic onset, the disease disseminates to involve lymph nodes and internal organs with a usually rapid fatal course.

In early states of the disease (stage IA, IB) the conventional treatment strategy is characterized by the use of topical therapy with increasing state-dependent aggressivity. This includes topical steroids, UV radiation, and topical cytotoxic substances such as nitrogen mustard. As CTCL progresses total body electron beam radiation therapy, X-irradiation, and, finally, systemic polychemotherapy is used. However, there are no data available which clearly document that any of these therapies can induce permanent remission; also the long-term benefit of nonstage-dependent early aggressive polychemotherapy has remained a matter of debate.

Natural sunlight has been known to clear skin lesions in early stages of mycosis fungoides, and this observation encouraged prospective studies of the effect of photochemotherapy in this disease.

The first encouraging results appeared in 1976.[31,32] Eight out of nine patients with eczematous or plaque stages (IA, IB) reported by Gilchrest et al. showed complete or at least 95% clearing.[31] In an Austrian series clearing was achieved in 12 out of 13 patients with stages IA or IB.[32] Since then several other investigators from both the U.S. and Europe have obtained similar favorable results.[33-36]

Treatment guidelines and dosimetry in oral photochemotherapy of CTCL are the same as established for the treatment of psoriasis. 8-MOP is the only furocoumarin tested so far, but other compounds may also prove to be effective. The treatment schedule is divided into a clearing phase, a maintenance phase, and a follow-up phase without therapy. As in psoriasis the initial phase of therapy consists of treatments four times weekly until complete resolution of skin lesions is achieved. Remission should be confirmed by histological examination of previously involved skin sites. This is followed by the maintenance phase consisting of two exposures per week for 1 month and one exposure per week for another month. If the patient is then still in remission, therapy is discontinued and the patient is monitored monthly or bimonthly. If a relapse occurs the patient is subjected to a new course of four PUVA exposures per week until complete clearing is achieved again.

Present experience indicates that CTCL responds dramatically to PUVA in stage IA, IB, and IIB where complete and long-lasting remissions can be induced. In an Austrian series of 44 patients,[47] the incidence of relapses paralleled the stage of the disease at which PUVA was started: patients who were in stage IA at the beginning of therapy experienced a lower incidence of relapses and a longer mean disease-free interval than patients in stage IB.[47] CTCL does not develop resistance against PUVA therapy, because relapses of eczematous and plaque lesions respond as well as the original lesions when photochemotherapy is resumed. Patients who relapse repeatedly may require a prolonged maintenance therapy.

Flat tumors in tumor-stage (IIB) patients may also eventually resolve, but this occurs only after long treatment periods and with high cumulative UVA radiations.[47] Moreover, most patients with tumors show a high rate of early relapses which can be barely controlled by permanent maintenance treatment, and only the combination of PUVA with local X-ray treatment and/or systemic polychemotherapy can induce complete tumor resolution.[47]

The majority of patients with early disease (IA, IB) can be kept in remission with maintenance treatment for up to 3 to 4.5 years.[39] In a recent follow-up study covering a mean observation period of 44 months, 56% of stage IA patients and 38% of stage IB patients were reported to have remained free of lesions after one single PUVA course.[47] Among patients who were monitored for more than 5 years, 75% of stage IA and 29% of stage IB patients had stayed in remission. In contrast, all patients with tumors experienced multiple relapses and only three out of seven survived a 6-year period despite combination treatment with ionizing radiation and/or polychemotherapy.[47] Obviously PUVA does not provide for lasting remissions once the disease has disseminated, but it may act synergistically with other therapeutic measures by reducing the tumor cell burden in the skin and thus may help to temporarily improve the patient's condition. The effect of such a combined treatment on patient survival is unknown. Possible long-term hazards related to frequent PUVA treatments are probably meaningless for patients with a malignant lymphoma as compared to patients with benign conditions, such as psoriasis.

Presently no therapeutic regimen is known to arrest the natural course of CTCL and to prevent progression to tumor formation and final fatal dissemination. Extracorporeal photopheresis, as discussed in the following section, appears to be a novel, promising therapeutic approach. With regard to the concept of clonal origin of the T-cell proliferation in CTCL, advocates of a primary aggressive treatment emphasize that polychemotherapy at the earliest possible stage could be curative in some cases, but this has not been investigated.

Photochemotherapy is now accepted as one of the most potent treatment forms for early stage CTCL. While PUVA therapy is very effective in inducing remissions as long as the lymphoma is confined to the skin, the effect of this therapy on the natural course of this disease and on patient survival is unknown.

E. Treatment of Vitiligo

Vitiligo is a pigmentary disorder of unknown etiology which is characterized by the loss of melanocytes and pigment from circumscribed areas of otherwise normal skin. The disease is neither dangerous nor life threatening, but it may be severely disfiguring and socially devastating, in particular in dark-skinned subjects. Treatment is necessary only when the disease causes considerable emotional and social distress.

Photochemotherapy stimulates melanogenesis and melanocyte proliferation and migration, and has been shown to be an effective therapeutic modality for vitiligo where it reconstitutes the normal color of depigmented skin in more than 50% of the patients.[48,49]

Oral 8-MOP or 4,5'8-methylpsoralen are the photosensitizers most frequently used followed by exposure to sunlight (PUVASOL) or to artificial UVA radiation. Based on a large controlled clinical trial comprising 365 patients by Pathak,[51] detailed information has now accumulated on different treatment schedules and on parameters to predict the efficacy of the treatment.

To induce repigmentation, vitiligo patients need constant long-term therapy (12 to 24 months). Because oral TMP is much less phototoxic than 8-MOP, TMP is the preferred drug for treatment with sunlight as radiation source. Up to 70% of vitiligo patients improved in the series of Mosher et al. when treated at least twice weekly for more than 1 year with oral TMP in a dose of 0.6 to 0.9 mg/kg.[51] Failure of repigmentation will require dose increments of TMP, a switch to 8-MOP, or a combination of both analogs. Treatments should be given at least twice a week; not more than three exposures a week are recommended, with at least 1 day between each treatment.[51] Failure to respond after approximately 25 treatments should be considered as unresponsiveness, and PUVA should then be terminated.

If the treatment is discontinued, reversal of the acquired repigmentation may occur unless the macule has completely repigmented. Past experience indicates that completely repigmented areas can be stable for a decade or more without loss of skin color. Progression of the disease during treatment despite the appearance of repigmentation in earlier lesions is not uncommon. This phenomenon does not necessarily indicate poor treatment response but may be due to suboptimal dosage. On the average a course of PUVA treatment consists of at least 150 exposures. Responders may show significant improvement after about 50 exposures.[51] Many patients eventually develop significant repigmentation, but total reconstitution of normal skin color is unusual as some skin areas do not respond at all despite many months of continuous therapy. In particular the vermillion part of the lips, the distal dorsal hands, tips of fingers and toes, areas of bony prominences, palms, soles, and nipples are very refractory to the treatment. It is therefore advisable to exclude patients with only acrofacial and lip-tip involvement from PUVA therapy.

F. Prevention Treatment of Photodermatoses

Tolerance to sunlight can be induced in several conditions with increased photosensitivity by increasing melanin pigmentation of skin with PUVA.[52-56] The PUVA-induced tan possibly acts as a natural sunscreen that prevents the development of UV-induced diseases.[57]

In polymorphic light eruption, the most common photodermatosis, PUVA represents a very effective prophylactic treatment. In about 70% of the patients with this condition a 3- to 4-week PUVA course consisting of two to three treatments per week suffices to completely suppress the disease upon subsequent exposure to sunlight.[58]

The decisive factor for the successful prevention of photodermatoses is pigmentation and a thickening of the horny layer which are known to absorb and scatter UV radiation and thus prevent its penetration into deeper layers.[57] An alternative hypothesis relates to possible immune phenomena, but as yet there exist no convincing data in support of the involvement of immune reactions in both the development and the photodermatoses.

II. PHOTOPHERESIS: EXTRACORPOREAL IRRADIATION OF 8-MOP-CONTAINING BLOOD

In the search for new therapeutic applications of photoactivatable drugs such as 8-MOP, a new therapeutic modality — extracorporeal photopheresis — has been developed.[59] This therapeutic approach, which showed initial success in controlling diseases such as intractable CTCL with minimal side effects and discomfort to the patient, was developed at the Columbia Presbyterian Medical Center in New York.[60] Ongoing international clinical trials have supported these initial observations.

This new approach was theoretically designed not only for the management of diseases caused by malignant lymphocytes such as CTCL or chronic lymphocytic leukemia (CLL) but also for treatment of other disorders where lymphocyte-mediated pathology is involved, i.e., autoimmune diseases such as pemphigus vulgaris, systemic lupus erythematosus, rheumatoid arthritis, myasthenia gravis, severe neurodermatitis, and treatment of organ transplant recipients who have been previously sensitized to donor antigens.

In brief, the procedure involves passage of blood containing 8-MOP previously ingested by the patient so as to obtain maximum plasma concentrations at the time the procedure is initiated (minimum, 50 ng/mℓ at 2 hr) from one arm vein through a photopheresis machine and back. In the machine subsequent separation of plasma and the white blood cell fraction (buffy coat, nucleated cells) from the red cell fraction of nonnucleated cells takes place. The last is returned without further treatment to the patient.

The 8-MOP-containing plasma and buffy coat fraction is subsequently treated as it flows through an UVA exposure system before it is also returned to the patient through another arm vein. In this fashion an otherwise pharmacologically inactive drug such as 8-MOP can literally be turned on in the bloodstream to cause profound clinically and immunologically desirable effects on circulating lymphocytes while sparing other body tissues.

For diseases involving white blood cells, particularly lymphocytes, this form of treatment permits a form of selectivity not possible with other presently available forms of drug therapy.

Extracorporeal photopheresis, which is currently still quite experimental, has induced surprising response in selected patients with the erythrodermic form of CTCL. Current multicentric international clinical trials on selected diseases of autoimmune nature should further help elucidate the effectiveness of this novel therapeutic tool as well as expand our understanding of the underlying mechanisms involved.[61-72]

A. Advantages of the System

Extracorporeal photopheresis, presently still to be considered an experimental treatment modality, is attractive for three main reasons. First, as elucidated previously in this chapter, 8-MOP, a naturally occurring substance found in small quantity in a variety of common vegetables and fruits, including parsely, limes, and figs, has been proven to be relatively safe and very effective in the treatment of psoriasis patients. Here, a benign hyperproliferative disease of the skin can be controlled by ingestion of the drug and subsequent UVA radiation of the skin without clinically significant short-term side effects.[73,74]

When compared to all other systemically administered drugs which cannot specifically target diseases tissues, 8-MOP appears to present with very desirable properties; the effects of 8-MOP are strictly limited to tissue exposed to UVA and result from the short life of the photoactive form. Basically, the drug returns to an inactive form once it leaves the exposed field.

Second, nucleated cells, particularly lymphocytes, appear to be quite sensitive to the effects of photoactivated 8-MOP.[62,68,70,71,75] As evidenced by in vitro studies and confirmed by monoclonal marker studies in CTCL patients treated by photopheresis, depending on the concentrations of 8-MOP or intensity of UVA, T lymphocytes can either be killed or functionally restricted.[111]

Third, with the encouraging results obtained from the psoriasis trials, the positive clinical results on CTCL patients and the knowledge that UVA can penetrate selected plastics just as well as it penetrates window glass, the concept that photoactivatable drugs such as 8-MOP could be directly activated or switched on in an extracorporeal blood-flow system and thus be used to treat these as well as other diseases mediated by lymphocytes became an intriging therapeutic alternative.

B. The Photopheresis Instrument

The prototype of the instrument used for the initial international clinical trials integrated an initial discontinuous leuka/plasma-pheresis step with subsequent ultraviolet exposure in a single unit. After oral administration of 8-MOP (in a dose previously determined to produce a minimum of 50 ng/ml of drug in the plasma 2 hr after ingestion) and with the patient reclining in bed, heparinized blood is leukapheresed in six cycles through a continuously spinning centrifuge bowl. This permits removal of 240 ml of leukocyte-enriched blood, which is then pooled with 300 ml of plasma obtained during the same procedure from the patient and 200 ml of sterile normal saline, yielding an expected final hematocrit of approximately 6 + 1% and containing an expected 50% of the number of blood lymphocytes at the start of the leukapheresis. In order to activate the 8-MOP contained in the plasma and nucleated cell blood fraction, the total volume of 740 ml is then passed through a six-chambered sterile cassette and exposed to UVA radiation.

Each chamber is composed of an outer polycarbonate sheath opaque to UVA and an inner UVA-transparent acrylic tube surrounding the fluorescent UVA energy source. The thickness of the fluid pumped between these two tubes is 1.0 mm. Flow in each chamber is from bottom to top with shunts running between the top of each chamber to the bottom of the subsequent chamber.

The total volume of this single-use sterile cassette is 190 mℓ. To allow continuous recycling of the blood through the cassette an automatically reversible blood pump is incorporated into this UVA exposure system as well as temperature sensors that ensure that blood temperature not rise above 41°C. Following exposure to UVA where the average lymphocyte is exposed to approximately 2 J/cm^2, the entire volume is returned to the patient.[112]

Depending on the various protocols currently under study in multicenter trials, this form of therapy is on the average repeated on 2 successive days with 4-week intervals but can, in selected nonresponsive patients, be repeated for limited periods of time with only 5-day intervals between treatment sets. A second-generation apparatus already in use in some centers has the capacity of concurrently radiating the cells as they are being separated from the blood. Time of treatment can be shortened to 4 hr, radiation cassettes can be reused, and the possibility of radiating with UVB and UVC to extend the range of applicability of this form of treatment has been incorporated.

C. Initial Clinical and Laboratory Observations

The most complete data presently available are on the initial group of CTCL patients treated in six centers in the U.S. and Europe (Austria and Germany).

Using a panel of commercially available T-cell markers as well as others, such as BE2, a tumor-associated marker positive in a number of CTCL patients, evaluation was made on the impact of photopheresis on selected T-cell subpopulations. Cell-sorter analysis of subsets performed on patients' cells the day before treatment, the day after treatment, and after the prescribed 4-week intervals showed a clearly discernible pattern: more then 40% average decreases in percentages of cells with each phenotype analyzed (including the BE2 marker) occurred with nearly complete restitution of values over the subsequent 4-week intervals. These results suggested that on the one hand large numbers of all T-cell classes could be damaged and eliminated from circulation and that tissue reservoirs and/or renewal rates were sufficient to replace them in the blood within 3- and 4-week intervals. On the other hand the surprising clinical responses observed initially in four of the first five patients, two of whom had failed to respond to twice-weekly leukapheresis, has made us come to the tentative conclusion that the reinfusion of lymphocytes damaged in this way may be of particular importance with immunochemotherapeutic relevance not hitherto described elsewhere.[113]

Over 40 CTCL patients have been submitted to this form of therapy. The impressive results observed in some centers has made of this therapeutic alternative the treatment of choice in the erythrodermic, exfoliative form of this disease. Compared to standard polychemotherapeutic regimens and electron-beam radiation, extracorporeal photopheresis has, within the observed time frame of over 4 years, not been shown to have significant deleterious side effects. The common side effects associated with conventional chemotherapy, such as bone marrow suppression, hair loss, gastrointestinal erosions, nausea, and vomiting, were not encountered.

The first reports on treatment of T-cell-mediated diseases, such as pemphigus vulgaris, systemic lupus erythematosus, psoriatic arthropathy, myasthenia gravis, and chronic lymphocytic leukemia, have been encouraging. The critical observation that this form of therapy does not appear to cause exacerbation of diseases by, for example, triggering immune-complex formation but on the contrary seems to bring about reduction of disease-specific antibodies with improvement of the clinical picture is very exciting. Recent studies on murine systemic lupus erythematosus treated by this therapeutic modality appear to support our initial observations.[76] Still, studies on some long- and short-term side effects of the PUVA treatment modality merit caution on the expansion of the spectrum of diseases treatable by extracorporeal photopheresis.[77,78]

D. Conclusion

The rapid progress in the development of this new and exciting therapeutic alternative allows one to express hopes that new routes of treatment of disease are becoming available. The potential behind this form of treatment is enormous, but until long-term studies have been evaluated one must express extreme caution in the use of this therapeutic modality in diseases that can easily be controlled by other relatively safe, cost-effective, time-proven agents.

At this early time, two points alone appear well established. First, it is possible to activate a photosensitive agent in extracorporally routed blood and return the blood to a human being without producing readily recognizable, significant, short-term side effects. Second, such treatment modality can induce significant clinical responses in at least certain situations.

Broad-based investigation of the mechanisms underlying the observed responses is necessary. From the clinical and laboratory data already generated, it is our impression that far more is being accomplished in vivo than merely the inactivation and destruction of exposed nucleated cells, particularly lymphocytes.

Thus it seems likely that the response to this therapy is secondary, in large measure, to the reinfusion of damaged, slowly dying cells. Whether we are actually immunizing the host against idiotypic determinants of the T-cell receptor on the abnormal cells remains to be fully elucidated in experimental models now under study. If it is possible to augment a natural clonotypic immune response against undesirable sets of T lymphocytes in humans, extracorporeal photopheresis may be a particularly effective means to this end. In this technique, lymphocytes which have been injured die over a period of several days and are presumably removed by a reticuloendothelial system which has not been altered by chemotherapy. A host of other explanations both trivial and complex are being explored and merit further investigation.

Extracorporeal photopheresis is currently being evaluated in multiinstitutional trials for the treatment of T-cell-mediated diseases. In addition to the previously mentioned disease entities, the experience with the treatment of two iatrogenically produced immunologic problems appears to be particularly attractive.

One of these entities comprises individuals with end-stage renal disease who are commonly prepared for kidney transplants from living, related donors by receiving whole blood transfusions from the potential donor. These transfusions present donor tissue antigens to the potential kidney recipient and in a high percentage of the cases make the recipient more likely to immunologically tolerate the subsequently transplanted kidney. Unfortunately the potential recipients often become immunized to exactly the tissue antigens to which it had been hoped they would become tolerant, thereby making the transplantation unfeasible. Recent indirectly related experiments on rat models using PUVA to prolong renal allograft survival time give encouraging support to this concept.[79,80] The second group would contain those patients receiving intravenous monoclonal antibody therapy for an unrelated disease. Experience in the last few years has repeatedly shown that these individuals nearly always become immunized to the mouse proteins, compromising the safety and efficacy of these antibodies. The two situations appear to be particularly conducive to clinical trials of photopheresis as an immunosuppressive therapy, since it will be possible to initiate treatment either at the time of administration of the immunogen, at the start of the immunologic response, or after the immunologic reaction has been fully established.

With knowledge of the structure of the immunogen, the sequential study of the capacity of photopheresis to inhibit undesirable immunologic reactions appears promising. Should prevention of immunologic reactions of this type be possible, the potential size of patient populations that could eventually be treated this way could be considerable.

Among the drugs of potential value, 8-MOP appears to be merely the first of a series of photosensitive compounds to be activated in plasma and blood cells. Aminomethyl-trimethylpsoralen (AMT) is effective at lower concentrations than is 8-MOP and requires lower levels of UVA to be effective.[81] Furthermore, AMT conjugated covalently with insulin is selectively internalized and following excitation by UVA damages human lymphocytes which have been previously stimulated in vitro by a mitogen (PHA) to develop large numbers of cell membrane insulin receptors.[114]

Gilvocarcin-V appears to be an effective antitumor antibiotic excitable by low doses of visible light.[82] Monoclonal antibodies have been linked to porphyrin molecules which lead to cellular destruction following their excitation by visible light.[83]

The further prospects for extracorporeal photochemotherapy thus appear to be vast. Still, the science behind this new therapeutic alternative is just beginning to crystallize. Very careful, critical, and methodical approaches to its development and clinical applications must be developed if its early optimistic evaluation will attain its full medical potential.

III. POTENTIAL CARCINOGENICITY OF PSORALEN PHOTOTHERAPIES

All forms of phototherapy, with and without photosensitizing drugs, are known to have mutagenic and carcinogenic effects. There exists ample experimental evidence that these effects are based on photochemical damage to cellular DNA.[83-87] The photochemical reaction in photochemotherapy involves the interaction of the psoralen molecule and DNA, which leads to the formation of monofunctional and bifunctional adducts. Unfortunately, this mechanism appears to be responsible for both the carcinogenic risk and the therapeutic effect. In particular, psoralens which form bifunctional adducts (cross-links) are more mutagenic and more therapeutically effective than monofunctional compounds. Since 8-MOP is the most widely used psoralen in therapy, most information available concerns the effects of this compound.

8-MOP administration and subsequent UVA irradiation is carcinogenic in animals, but with the oral route the rate of skin tumors is lower than after topical or intraperitoneal application.[83,85,86,88] While there is no doubt about the carcinogenic potential of PUVA, the question arises whether the therapeutic benefits outweigh the risk of its long-term use. No increased incidence of skin carcinomas was detected in patients with vitiligo who had been treated in the past with oral trisoralen and sunlight for up to a decade.[89] However, these patients were never subjected to a comparably meticulous monitoring, as is done with psoriatics. Moreover, as opposed to psoriasis patients, patients with vitiligo have generally not received other potentially carcinogenic treatment forms prior to PUVA therapy.

In 1979 the American 16-center cooperative study was the first to report on an increased risk of nonmelanoma skin cancers in PUVA-treated patients, particularly in those who had been previously treated with ionizing radiation or who had had a history of skin carcinomas prior to PUVA.[90] A causal relationship with PUVA was suggested by the fact that many of these tumors had occurred on body areas not previously sun exposed. This increased risk became evident 2 years after the initiation of photochemotherapy, implicating a very short latency period.[90] Over a period of another 2 years no statistically significant increase in basal cell carcinomas was observed, but there was a dose-dependent increase in the incidence of squamous cell carcinomas.[91] The conclusions drawn in this study were not unequivocally accepted because no appropriate control population had been included[92] and because no increased incidence in skin carcinomas has been found in several European[93-97] and two other American[92,98] studies. However, in these studies a direct correlation between tumor development and certain risk factors, such as previous treatments with arsenic and/or ionizing radiation, was detected.[94-96,99] A similar observation was reported in a small group of patients with mycosis fungoides treated with nitrogen mustard and radiation therapy before PUVA who developed squamous cell carcinomas, keratoacanthomas, and keratoses.

The question remains to be clarified whether PUVA per se is oncogenic or acts as a promoter if certain genetic or other risk factors are present. The short latency period observed for the carcinomas encountered so far could be suggestive of a promoter rather than an initiator effect. In this context it should be noted that the biological behavior of these carcinomas can be classified as being low aggressive in that no metastasizing disease was reported. However, in some lesions it proved to be difficult to discriminate between true carcinomas, invasive actinic keratoses, and keratoacanthomas.

In 1984 the American 16-center follow-up study documented a 12.8-fold risk of squamous cell carcinoma in patients receiving a high-dosage therapy as opposed to patients with a low total exposure to PUVA. No such dose-dependent increase of risk was found for basal cell carcinomas.[100] In contrast with this experience, European patient cohorts so far do not exhibit such a trend, although the total PUVA exposure doses were comparable in some of the studies.[101] At present there exists no explanation for this discrepancy, but it may be important to note that European and American treatment protocols differ considerably. The 16-center study protocol[15] employs a more cautious approach with a large number of lower single doses as compared with the aggressive European schedule[17] which provided for fewer but higher single-dose treatments.

Current discussions now focus on the question of whether few but large phototoxic PUVA doses are more harmful than frequent small doses and whether an aggressive regimen with rest periods is safer than a continuous nonaggressive treatment.[102] A recent study which addressed the question of photocarcinogenesis with 8-MOP and UVA in mice has unequivocally demonstrated that the law of dose reciprocity does not hold for PUVA carcinogenesis: in terms of cumulative PUVA doses, the lower daily doses proved to be more carcinogenic than the higher daily doses.[100]

According to current knowledge it remains unsettled whether limits of PUVA dose can be established within which photochemotherapy can be considered safe for the treatment of benign dermatoses. In the treatment of malignant processes such as CTLC possible long-term side effects of PUVA may be less significant vis-á-vis more aggressive alternative treatment options (electron beam radiation therapy and polychemotherapy).[47] Current developments are now directed at reducing the total long-term PUVA dose. New strategies aiming at this goal include the combination of PUVA with other therapeutic measures to achieve a dose-saving effect. For psoriasis a combination with topical corticosteroids and anthralin increases the efficacy of PUVA,[22] and a combination with methotrexate has been used with success.[103] Chemophotochemotherapy (RePUVA), a combination of PUVA with retinoids, cuts the total PUVA dose required for clearing psoriasis by more than 50% in comparison to PUVA alone, and this is considered a decisive advance in therapy.[24,25,104] However, the photobiologic activity of psoralens and related compounds can be modified by changing one or more parts of the molecule. The aim of such a process is to develop new analogues with reduced cytogenetic hazards which still are effective in therapy. Until such new compounds or new treatment regimens are available, as an alternative patients with severe recurrent psoriasis could be cycled through several different treatment forms in order to remain below the critical toxic or carcinogenic doses of each treatment modality.[105]

There is also concern that PUVA may increase the risk of malignant melanoma; and some morphological changes suggest an irreversible degenerative damage to melanocytes. PUVA increases melanocytic activity; it stimulates repigmentation in vitiligo and induces lentigo-like freckles. However, at the time of writing only three melanomas[106-108] and two melanomas *in situ*[109] have been reported in PUVA-treated patients. These figures are well below the expected incidence of melanoma in the normal Caucasian population. Also there is no evidence for an increased risk of melanoma in all prospective PUVA studies published so far.

PUVA represents a highly effective form of therapy for a number of severe skin diseases. It can be considered safe if used as a short-term therapy provided that certain guidelines are carefully observed. Defined criteria for patient selection, dosimetry, and follow-up as mandatory in other forms of chemotherapy also apply for PUVA. The risk-benefit ratio must be considered when PUVA is utilized for prolonged treatment.

ACKNOWLEDGMENT

R. M. Knobler held a Max Kade Foundation fellowship while a visiting scientist at Columbia Presbyterian Medical Center in 1983 through 1985. This work was funded in part by National Institutes of Health grant Ca 20499 and by the Matheson Foundation.

REFERENCES

1. **Parrish, J. A., Fitzpatrick, T. B., Tanenbaum, L., and Pathak, M. A.,** Photochemotherapy of psoriasis with oral methoxsalen and long wave ultraviolet light, *N. Engl. J. Med.*, 291, 1207, 1974.
2. **Wolff, K., Hönigsmann, H., Gschnait, F., and Konrad, K.,** Photochemotherapie bei Psoriasis. Klinische Erfahrungen bei 152 Patienten, *Dtsch. Med. Wochenschr.*, 100, 2471, 1975.
3. **Wolff, K., Fitzpatrick, T. B., Parrish, J. A., Gschnait, F., Gilchrest, B. A., Hönigsmann, H., Pathak, M. A., and Tanenbaum, L.,** Photochemotherapy with orally administered methoxsalen, *Arch. Dermatol.*, 112, 943, 1976.
4. **El Mofty, A. M.,** A perliminary clinical report on the treatment of leukoderma with *Ammi majus* Linn., *J. R. Egypt. Med. Assoc.*, 31, 651, 1948.
5. **Wolff, K., Gschnait, F., Hönigsmann, H., Konrad, K., Parrish, J. A., and Fitzpatrick, T. B.,** Phototesting and dosimetry for photochemotherapy, *Br. J. Dermatol.*, 96, 1, 1977.
6. **Morison, W. L.,** *Phototherapy and Photochemotherapy of Skin Disease*, Praeger, New York, 1983.
7. **Wolff, K. and Hönigsmann, H.,** Clinical aspects of photochemotherapy, in *International Encyclopedia of Pharmacology and Therapeutics*, Section 110, *The Chemotherapy of Psoriasis*, Baden, H. P., Ed., Pergamon Press, Oxford, 1984, 247.
8. **Hönigsmann, H. and Stingl, G., Eds.,** *Current Problems in Dermatology*, Vol 15, *Therapeutic Photomedicine*, S. Karger Basel, 1986.
9. **Hönigsmann, H., Wolff, K., Fitzpatrick, T. B., Pathak, M. A., and Parrish, J. A.,** Oral photochemotherapy with psoralens and UVA (PUVA): principles and practice, in *Dermatology in General Medicine*, 3rd ed., Fitzpatrick, T. B., Eisen, A. Z., Wolff, K., Freedberg, I. M., and Austen, F. K., Eds., McGraw-Hill, New York, 1986.
10. **Hönigsmann, H., Jaschke, E., Gschnait, F., Brenner, W., Fritsch, P. O., and Wolff, K.,** 5-Methoxypsoralen (Bergapten) in photochemotherapy of psoriasis, *Br. J. Dermatol.*, 101, 369, 1979.
11. **Hönigsmann, H.,** Psoralen photochemotherapy — mechanisms, drugs, toxicity, in *Current Problems in Dermatology*, Vol. 15, *Therapeutic Photomedicine*, Hönigsmann, H. and Stingl. G, Eds., S. Karger, Basel, 1986, 52.
12. **Hönigsmann, H., Jaschke, E., Nitsche, V., Brenner, W., Rauschmeier, W., and Wolff, K.,** Serum levels of 8-methoxypsoralen in two different drug preparations. Correlation with photosensitivity and UV-A dose requirements for photochemotherapy, *J. Invest. Dermatol.*, 79, 233, 1982.
13. **Pathak, M. A.,** Mechanisms of psoralen photosensitization and in vivo biological action spectrum of 8-methoxypsoralen, *J. Invest. Dermatol.*, 37, 397, 1961.
14. **Young, A. R. and Magnus, I. A.,** An action spectrum for 8-MOP-induced sunburn cells in mammalian epidermis, *Br. J. Dermatol.*, 104, 541, 1981.
15. **Melski, J. A., Tanenbaum, L., Parrish, J. A., Fitzpatrick, T. B., and Bleich, H. L.,** Oral methoxsalen photochemotherapy for the treatment of psoriasis: a cooperative clinical trial, *J. Invest. Dermatol.*, 68, 328, 1977.
16. Clinical cooperative study of PUVA-48 and PUVA-64: photochemotherapy of psoriasis, *Arch. Dermatol.*, 115, 576, 1979.

17. **Henseler, T., Wolff, K., Hönigsmann, H., and Christophers, E.,** The European PUVA study (EPS): oral 8-methoxypsoralen photochemotherapy of psoriasis. A cooperative study among 18 European centers, *Lancet,* 1, 853, 1981.

18. **Hönigsmann, H., Gschnait, F., Konrad, K., and Wolff, K.,** Photochemotherapy for pustular psoriasis (von Zumbusch), *Br. J. Dermatol.,* 97, 119, 1977.

19. **Morison, W. L., Parrish, J. A., and Fitzpatrick, T. B.,** Oral methoxsalen photochemotherapy of recalcitrant dermatoses of palms and soles, *Br. J. Dermatol.,* 99, 297, 1978.

20. **Murray, D., Corbett, M. F., and Warin, A. P.,** A controlled trial of photochemotherapy for persistent palmoplantar pustulosis, *Br. J. Dermatol.,* 102, 659, 1980.

21. **Schmoll, M., Henseler, T., and Christophers, E.,** Evaluation of PUVA, topical corticosteroids, and the combination of both in the treatment of psoriasis, *Br. J. Dermatol.,* 99, 693, 1978.

22. **Morison, W. L., Parrish, J. A., and Fitzpatrick, T. B.,** Controlled study of PUVA and adjunctive topical therapy in the management of psoriasis, *Br. J. Dermatol.,* 98, 125, 1978.

23. **Momtaz, K. and Parrish, J. A.,** Combination UVB and PUVA in the treatment of psoriasis, *J. Invest. Dermatol.,* 76, (Abstr.), 303, 1981.

24. **Fritsch, P. O., Hönigsmann, H., Jaschke, E., and Wolff, K.,** Augmentation of oral methoxsalen-photochemotherapy with an oral retinoic acid derivative, *J. Invest. Dermatol.,* 70, 178, 1978.

25. **Hönigsmann, H. and Wolff, K.,** Isotretinoin-PUVA for psoriasis, *Lancet,* 1, 236, 1983.

26. **Heidbreder, G. and Christophers, E.,** Therapy of psoriasis with retinoid plus PUVA: clinical and histologic data, *Arch. Dermatol. Res.,* 264, 331, 1979.

27. **Grupper, C. and Berretti, B.,** Treatment of psoriasis by oral PUVA therapy combined with aromatic retinoid (Ro 10-9359; Tigason®), *Dermatologica,* 162, 404, 1981.

28. **Laurahanta, J., Juvakoski, T., and Lassus, A.,** A clinical evaluation of the effects of an aromatic retinoid (Tigason), combination of retinoid and PUVA, and PUVA alone in severe psoriasis, *Br. J. Dermatol.,* 104, 325, 1981.

29. **Morison, W. L., Momtaz, K., Parrish, J. A., and Fitzpatrick, T. B.,** Combined methotrexate-PUVA in the treatment of psoriasis, *J. Am. Acad. Dermatol.,* 6, 46, 1982.

30. **Fitzsimons, C. P., Long, J., and MacKie, R.,** Synergistic carcinogenic potential of methotrexate and PUVA in psoriasis, *Lancet,* 1, 235, 1983.

31. **Gilchrest, B. A., Parrish, J. A., Tanenbaum, L., Haynes, H. A., and Fitzpatrick, T. B.,** Oral methoxsalen photochemotherapy of mycosis fungoides, *Cancer,* 38, 683, 1976.

32. **Hönigsmann, H., Konrad, K., Gschnait, F., and Wolff, K.,** Photochemotherapy of mycosis fungoides (Abstr.), in *7th Int. Cong. Photobiology,* Rome, 1976, 222.

33. **Roenigk, H. H., Jr.,** Photochemotherapy for mycosis fungoides, *Arch. Dermatol.,* 113, 1047, 1977.

34. **Hofmann, C., Burg, G., Plewig, G., and Braun-Falco, O.,** Photochemotherapie kutaner Lymphome. Orale und lokale 8-MOP-UVA Therapie, *Dtsch. Med. Wochenschr.,* 102, 675, 1977.

35. **Konrad, K., Gschnait, F., Hönigsmann, H., Fritsch, P., and Wolff, K.,** Photochemotherapie bei Mycosis fungoides, *Hautarzt,* 29, 191, 1978.

36. **Bleehen, S. S., Vella Briffa, D., and Warin, A. P.,** Photochemotherapy in mycosis fungoides, *Clin. Exp. Dermatol.,* 3, 377, 1978.

37. **Lowe, N. J., Cripps, D. J., Dufton, P. A., and Vickers, C. F. H.,** Photochemotherapy for mycosis fungoides, *Arch. Dermatol.,* 115, 50, 1979.

38. **Roenigk, H. H., Jr.,** Photochemotherapy for mycosis fungoides. Long-term follow up study, *Cancer Treat. Rep.,* 63, 669, 1979.

39. **Gilchrest, B. A.,** Methoxsalen photochemotherapy for mycosis fungoides, *Cancer Treat. Rep.,* 63, 663, 1979.

40. **Molin, L., Thomsen, K., Voldén, G., and Groth, O.,** Photochemotherapy (PUVA) in the pretumour stage of mycosis fungoides: a report from the Scandinavian mycosis fungoides study group, *Acta Derm. Venereol. (Stockholm),* 61, 47, 1980.

41. **Vella Briffa, D., Warin, A. P., Harrington, C. I., and Bleehen, S. S.,** Photochemotherapy in mycosis fungoides. A study of 73 patients, *Lancet,* 2, 49, 1980.

42. **Oberste-Lehn, H.,** Photochemotherapie der T-Zellen-Lymphome, *Z. Hautkr.,* 55, 1335, 1980.

43. **Abel, E. A, Deneau, D. G., Farber, E. M., Price, N. M., and Hopper, R. T.,** PUVA treatment of erythrodermic and plaque type mycosis fungoides, *J. Am. Acad. Dermatol.,* 4, 423, 1981.

44. **Molin, L., Thomsen, K., Voldén, G., and Groth, O.,** Photochemotherapy (PUVA) in the tumour stage of mycosis fungoides: a report from the Scandinavian mycosis fungoides study group, *Acta Derm. Venereol. (Stockholm),* 61, 52, 1981.

45. **Warin, A. P.,** Photochemotherapy in the treatment of psoriasis and mycosis fungoides, *Clin. Exp. Dermatol.,* 6, 651, 1981.

46. **Hamminga, L., Hermans, J., Noordijk, E. M., Meijer, C. J. L. M., Scheffer, E., and van Vloten, W. A.,** Cutaneous T-cell lymphoma: clinicopathological relationships, therapy and survival in ninety-two patients, *Br. J. Dermatol.,* 107, 145, 1982.

47. **Hönigsmann, H., Brenner, W., Rauschmeier, W., Konrad, K., and Wolff, K.,** Photochemotherapy for cutaneous T cell lymphoma. *J. Am. Acad. Dermatol.,* 10, 238, 1984.
48. **El Mofty, A. M.,** *Vitiligo and Psoralens,* Pergamon Press, Oxford, 1968.
49. **Parrish, J. A., Fitzpatrick, T. B., Shea, C., and Pathak, M. A.,** Photochemotherapy of vitiligo, *Arch. Dermatol.,* 112, 1531, 1976.
50. **Pathak, M. A., Mosher, D. B., Fitzpatrick, T. B., and Parrish, J. A.,** Relative effectiveness of three psoralens and sunlight in repigmentation of 365 vitiligo patients, *J. Invest. Dermatol.,* 74, (Abstr.), 252, 1980.
51. **Mosher, D. B., Pathak, M. A., and Fitzpatrick, T. B.,** Vitiligo: etiology, pathogenesis, diagnosis and treatment, in *Dermatology in General Medicine: Update One,* Fitzpatrick, T. B., Eisen, A. Z., Wolff, K., Freedberg, I. M., and Austen, F. K., Eds., McGraw-Hill, New York, 1983, 205.
52. **Gschnait, F., Hönigsmann, H., Brenner, W., Fritsch, P., and Wolff, K.,** Induction of UV light tolerance by PUVA in patients with polymorphous light eruption, *Br. J. Dermatol.,* 99, 293, 1978.
53. **Parrish, J. A., LeVine, M. J., Morison, W. L., Gonzalez, E., and Fitzpatrick, T. B.,** Comparison of PUVA and beta-carotene in the treatment of polymorphous light eruption, *Br. J. Dermatol.,* 100, 187, 1979.
54. **Hölzle, E., Hofmann, C., and Plewig, G.,** PUVA treatment for solar urticaria and persistent light reaction, *Arch. Dermatol. Res.,* 269, 87, 1980.
55. **Jaschke, E. and Hönigsmann, H.,** Hydroa vacciniforme — Aktionsspektrum, UV-Toleranz nach Photochemotherapie, *Hautarzt,* 32, 350, 1981.
56. **Parrish, J. A., Jaenicke, K. F., Morison, W. L., Momtaz, K., and Shea, C.,** Solar urticaria: treatment with PUVA and mediator inhibitors, *Br. J. Dermatol.,* 106, 575, 1982.
57. **Gschnait, F., Brenner, W., and Wolff, K.,** Photoprotective effect of a psoralen-UVA-induced tan, *Arch. Dermatol. Res.,* 263, 181, 1978.
58. **Ortel, B., Tanew, A., Wolff, K., and Hönigsmann, H.,** Polymorphous light eruption. Action spectrum and photoprotection, *J. Am. Acad. Dermatol.,* 14, 748, 1986.
59. **Edelson, R., Berger, C., Gasparro, F., Lee, K. E., and Taylor, T.,** Treatment of leukemic cutaneous T cell lymphoma with extracorporeally-photoactivated 8-methoxypsoralen, *Clin. Res.,* 31, 467A, 1983.
60. **Knobler, R. M. and Edelson, R. L.,** Cutaneous T cell lymphoma in the medical clinics of North America, in *Cutaneous Oncology,* Vol. 70, Callen, J. P. and Allegra, J., Eds., W. B. Saunders, Philadelphia, 1986, 109.
61. **Binet, J. L., Villeneuve, B., Van Rapenbusch, R., Mignon, F., Becard, R., Vaughier, G., and Bernard, J.,** L'irradiation extra-corporalle du sang par les rayons ultraviolets, *Nouv. Rev. Hematol.,* 8, 733, 1968.
62. **Krüger, J. P., Christophers, E., and Schlaak, M.,** Dose-effects of 8-methoxypsoralen and UVA in cultured human lymphocytes, *Br. J. Dermatol.,* 98, 141, 1978.
63. **Gasparro, F. P., Berger, C. L., and Edelson, R. L.,** Effect of monochromatic UVA light and 8-methoxypsoralen on human lymphocyte response to mitogen, *Photodermatology,* 1, 10, 1984.
64. **Gunn, A., Scrimgeour, D., Potts, R. C., Mackenzie, L. A., Brown, R. A., and Swanson-Beck, J.,** The destruction of peripheral blood lymphocytes by extracorporeal exposure to ultraviolet radiation, *Immunology,* 50, 477, 1983.
65. **Edelson, R. L.,** Extracorporeal photopheresis, *Photodermatology,* 1, 209, 1984.
66. **Gasparro, F. P., Song, J., Knobler, R. M., and Edelson, R. L.,** Quantitation of psoralen photoadducts in DNA isolated from lymphocytes treated with 8-methoxypsoralen and ultraviolet A radiation (extracorporeal photopheresis), in *Current Problems in Dermatology,* Vol. 15, Therapeutic Photomedicine, Hönigsmann, H. and Stingl, G., Eds., S. Karger, Basel, 1986, 67.
67. **Murina, M. A. and Roshchupkin, D. I.,** Effects of ultraviolet radiation on aggregation of human erythrocytes, *Photobiochem. Photobiophys.,* 7, 59, 1984.
68. **Bredberg, A. and Forsgren, A.,** Effects of in vitro PUVA on human leukocyte function, *Br. J. Dermatol.,* 111, 159, 1984.
69. **Hanson, C. V., Riggs, J. L., and Lennette, E. H.,** Photochemical inactivation of DNA and RNA viruses by psoralen derivatives, *J. Genet. Virol.,* 40, 345, 1978.
70. **Kraemer, K. H., Waters, H. L., Cohen, L. F., Popescu, N. C., Amsbaugh, S. C., DiPaolo, J. A., Glaubiger, D., Ellington, L. O., and Tarone, E. R.,** Effects of 8-methoxypsoralen and ultraviolet radiation on human lymphoid cells in vitro, *J. Invest. Dermatol.,* 76, 80, 1981.
71. **VanDuin, M., Westerveld, A., and Hoeijmakers, J. H. J.,** UV stimulation of DNA-mediated transformation of human cells, *Mol. Cell. Biol.,* 5, 734, 1985.
72. **Krylenkov, V. A., Kukui, L. M., Malygin, A. M., Osmanov, M. A., Shaphin, A. G, and Kholmogorov, V. E.,** Ultraviolet irradiation of blood: photochemistry, immunologic effect, *DNA,* 270, 242, 1983.
73. **Parrish, J. A., Fitzpatrick, T. B., Pathak, M. A., and Tannenbaum, L.,** Photochemotherapy of psoriasis with oral methoxsalen and longwave ultraviolet light, *N. Engl. J. Med.,* 291, 1207, 1974.

74. **Wolff, K., Fitzpatrick, T. B., and Parrish, J. A.,** Photochemotherapy for psoriasis with orally administered methoxypsoralen, *Arch. Dermatol.,* 112, 943, 1976.

75. **Kraemer, K. H., Waters, H. L., Ellingson, O. L., and Tarone, R. E.,** Psoralen plus ultraviolet radiation-induced inhibition of DNA synthesis and viability in human lymphoid cells in vitro, *Photochem. Photobiol.,* 30, 263, 1979.

76. **Perez, M., Gapas, Y., O'Neil, D., Edelson, R. L., and Berger, C. L.,** Inhibition of murine autoimmune disease by re-infusion of syngeneic lymphocytes inactivated with psoralen and ultraviolet A light, *J. Invest. Dermatol.,* 86, 494, 1986.

77. **Faed, M. J. W., Williamson, L., Peterson, S., Lakshipathi, I., Johnson, B. E., and Frain-Bell, W.,** Sister chromatid exchange and chromosome aberration rates in a group of psoriatics before and after a course of PUVA treatment, *Br. J. Dermatol.,* 102, 295, 1980.

78. **Wolff-Schreiner, E., Calzer, M., Schwarzacher, H. G., and Wolff, K.,** Sister chromatid exchanges in photochemotherapy, *J. Invest. Dermatol.,* 69, 387, 1977.

79. **Osterwitz, H., Scholz, D., Kaden, J., and Mebel, M.,** Prolongation of rat renal allograft survival time by donor pretreatment with 8-methoxypsoralen and longwave ultraviolet irradiation of the graft (PUVA therapy), *Urol. Res.,* 13, 95, 1985.

80. **Osterwitz, H., Kaden, J., Scholz, D., Mrochen, H., and Mebel, M.,** Synergistic effect of donor pretreatment with 8-methoxypsoralen and ultraviolet irradiation of the graft plus azathioprine and prednisolone therapy in prolonging rat renal allograft survival, *Urol. Res.,* 14, 21, 1986.

81. **Gasparro, F. P., Song, J., Knobler, R. M., and Edelson, R. L.,** Quantitation of psoralen photoadducts in DNA isolated from lymphocytes treated with 8-methoxypsoralen and ultraviolet A radiation (extracorporeal photopheresis), in *Current Problems in Dermatology,* Vol. 15, *Therapeutic Photomedicine,* Hönigsmann, H. and Stingl. G., Eds., S. Karger, Basel, 1986, 67.

82. **Knobler, R. M., Lane, M. J., Gasparro, F. P., Saffran, W. A., Morse, R., and Köck, A.,** DNA sequence specificity of gilvocarcin V, a new photoactivatable drug, *J..Invest. Dermatol.,* 86, 331, 1986.

83. **Oseroff, A. R., Wimberley, T., Lee, C., Alvarez, C., and Parrish, T. A.,** Photosensitized destruction of normal and leukemic T cells using monoclonal antibody directed hematoporphyrin, *J. Invest. Dermatol.,* 84, 335A, 1985.

84. **Griffin, A. G.,** Methoxsalen in ultraviolet carcinogenesis in the mouse, *J. Invest. Dermatol.,* 32, 367, 1959.

85. **Igali, S., Bridges, B. F., Ashwood-Smith, M. J., and Scott, B. R.,** Mutagenesis in *Escherichia coli,* *Mutat. Res.,* 9, 21, 1970.

86. **Langner, A., Wolska, H., Marzulli, F. N., Jablonska, S., and Jarzabek-Chorzelska, M.,** Dermal toxicity of 8-methoxypsoralen administered (by gavage) to hairless mice irradiated with longwave ultraviolet light, *J. Invest. Dermatol.,* 69, 451, 1977.

87. **Pathak, M. A., Daniels, F., Jr., Hopkins, C. E., and Fitzpatrick, T. B.,** Ultraviolet carcinogenesis in albino and pigmented mice receiving furocoumarins: psoralen and 8-methoxypsoralen, *Nature (London),* 183, 728, 1959.

88. **Schenley, R. L. and Hsie, A. W.,** Interaction of 8-methoxypsoralen and near-UV light causes mutation and cytotoxicity in mammalian cells, *Photochem. Photobiol.,* 33, 179, 1981.

89. **Urbach, F.,** Modification of ultraviolet carcinogenesis by photoactive agents: preliminary report, *J. Invest. Dermatol.,* 32, 373, 1959.

90. **Fitzpatrick, T. B., Parrish, J. A., Pathak, M. A., and Tanenbaum, L.,** The risks and benefits of oral PUVA photochemotherapy of psoriasis, in *Psoriasis,* Farber, E. M. and Cox, A. J., Eds., Yorke Medical Books, New York, 1977, 320.

91. **Stern, R. S., Thibodeau, L. A., Kleinerman, R. A., Parrish, J. A., and Fitzpatrick, T. B.,** Risk of cutaneous carcinoma in patients treated with oral methoxsalen photochemotherapy for psoriasis, *N. Engl. J. Med.,* 300, 809, 1979.

92. **Stern, R. S., Parrish, J. A., Bleich, H. L., and Fitzpatrick, T. B.,** PUVA (psoralen and ultraviolet A) and squamous cell carcinoma in patients with psoriasis, *J. Invest. Dermatol.,* 76, (Abstr.), 311, 1981.

93. **Halprin, K. M.,** Psoriasis, skin cancer, and PUVA, *J. Am. Acad. Dermatol.,* 2, 334, 1980.

94. **Grupper, C. and Berretti, B.,** Tar, ultraviolet light, PUVA and cancer, *J. Am. Acad. Dermatol.,* 3, 643, 1980.

95. **Hönigsmann, H., Wolff, K., Gschnait, F., Brenner, W., and Jaschke, E.,** Keratoses and non-melanoma skin tumors in long-term photochemotherapy (PUVA), *J. Am. Acad. Dermatol.,* 3, 406, 1980.

96. **Lassus, A., Reunala, T., Idänpään-Heikkilä, J., Juvakoski, T., and Salo, O. P.,** PUVA treatment and skin cancer. A followup study, *Acta Dermatol. Venereol. (Stockholm),* 61, 141, 1981.

97. **Lindskov, R.,** Skin carcinomas and treatment with photochemotherapy (PUVA), *Acta Dermatol. Venereol. (Stockholm),* 63, 223, 1983.

98. **Ros, S.-M., Wennersten, G., and Lagerholm, B.,** Long-term photochemotherapy for psoriasis: a histopathological and clinical follow-up study with special emphasis on tumour incidence and behaviour of pigmented lesions, *Acta Dermatol. Venereol. (Stockholm),* 63, 215, 1983.

99. **Roenigk, H. H., Jr. and Caro, W. A.**, Skin cancer in the PUVA-48 cooperative study, *J. Am. Acad. Dermatol.*, 4, 319, 1981.

100. **Reshad, H., Challoner, F., Pollock, D. J., and Baker, H.**, Cutaneous carcinoma in psoriatic patients treated with PUVA, *Br. J. Dermatol.*, 110, 299, 1984.

101. **Gibbs, N. K., Young, A. R., and Magnus, I. A.**, Failure of UVR dose reciprocity for skin tumorigenesis in hairless mice treated with 8-methoxypsoralen, *Photochem. Photobiol.*, 42, 39, 1985.

102. **Tanew, A., Hönigsmann, H., Hammerschmied, W., Rauschmeier, W., and Wolff, K.**, Keratoses and non-melanoma skin tumors in long-term methoxsalen photochemotherapy. A follow-up study, *Photochem. Photobiol.*, 39 (Suppl.), 60S, 1984.

103. **Gibbs, N. K., Hönigsmann, H., and Young, A. R.**, PUVA treatment strategies and cancer risk, *Lancet*, 1, 150, 1986.

104. **Morison, W. L., Momtaz, K., Parrish, J. A., and Fitzpatrick, T. B.**, Combined methotrexate-PUVA in the treatment of psoriasis, *J. Am. Acad. Dermatol.*, 6, 46, 1982.

105. **Fritsch, P. O. and Hönigsmann, H.**, Combination phototherapy. A critical appraisal, in *Current Problems in Dermatology*, Vol. 15, *Therapeutic Photomedicine*, Hönigsmann, H. and Stingl, G., Eds., S. Karger, Basel, 1986, 238.

106. **Parrish, J. A., Stern, R. S., Pathak, M. A., and Fitzpatrick, T. B.**, Photochemotherapy of skin diseases, in *The Science of Photomedicine*, Regan, J. D. and Parrish, J. A., Eds., Plenum Press, New York, 1982, 595.

107. **Forrest, J. B. and Forrest, J. H.**, Malignant melanoma arising during drug therapy for vitiligo, *J. Surg. Oncol.*, 12, 337, 1980.

108. **Frenk, E.**, Malignant melanoma in a patient with severe psoriasis treated by oral methoxsalen photochemotherapy, *Dermatologica*, 167, 152, 1983.

109. **Kemmet, D., Reshad, H., and Baker, H.**, Nodular malignant melanoma and multiple squamous cell carcinomas in a patient treated by photochemotherapy for psoriasis, *Br. Med. J.*, 289, 498, 1984.

110. **Marx, J. L., Auerbach, R., Possick, P., Myrow, R., Gladstein, A. H., and Kopf, A. W.**, Malignant melanoma in situ in two patients treated with psoralens and ultraviolet A, *J. Am. Acad. Dermatol.*, 9, 904, 1983.

111. unpublished.

112. in press.

113. **Edelson, R. L.**, in press.

114. in press.

Chapter 9

ROLE OF PSORALEN RECEPTORS IN CELL GROWTH REGULATION

Jeffrey D. Laskin and Debra L. Laskin

TABLE OF CONTENTS

I. INTRODUCTION

Over the last several years evidence has accumulated to support the model that cell growth is regulated by hormones and growth factors. A number of these growth factors have been characterized, including epidermal growth factor (EGF), fibroblast growth factor, platelet-derived growth factor, and transforming growth factors.[1-11] The growth regulatory properties of steroid hormones, prostaglandins, and their metabolites, as well as the retinoids, have also been the subject of intensive investigation.[12-24] A common property of these compounds is that they mediate their biological actions by binding to or modulating specific high-affinity receptor sites located on target cells. The precise biochemical events leading to alterations in cell growth following activation of these receptors is varied and depends on the nature of the growth regulatory substance and on the target cell.

Several model xenobiotics are also potent growth regulatory substances. Two of the best characterized of these chemicals are the polychlorinated polycyclic aromatic hydrocarbon, 2,3,7,8-tetrachlorodibenzo-*p*-dioxin (TCDD) and the phorbol ester, 12-*O*-tetradecanoyl-phorbol-13-acetate (TPA). TCDD is one of the most toxic synthetic chemicals known.[25-33] Among the many biological effects of this toxin in humans is the development of epidermal hyperkeratinization which may culminate in chloracne, a characteristic persistent skin condition.[34,35] TPA is a potent skin irritant and is a well-known tumor-promoting agent.[36,37] In the mouse model, both TCDD and TPA induce epidermal hyperplasia and ornithine decarboxylase, and alter patterns of keratinization.[27,33,38-40] TCDD, like TPA, is also a tumor promoter in the two-stage mouse skin model for carcinogenesis.[41] Current evidence suggests that distinct receptor proteins mediate the biological effects of TCDD and TPA.[28,42,43] Thus, like the growth factors described above, these compounds have highly specific cellular targets. TPA and TCDD binding in sensitive cells followed by receptor activation initiates a cascade of intracellular events leading to the diverse biological effects observed with these compounds. Recently, it has been shown that receptor activation by these toxins can also lead to alterations in normal growth factor receptors. For example, both TPA and TCDD can inhibit EGF receptor binding to cells, although their mechanisms are distinct.[44-52] Independent of the mechanism, however, the interaction of these compounds with the EGF receptor may partially explain the action of these toxins.

Our laboratory has been interested in another class of biologically active xenobiotics known as furocoumarins or psoralens. These compounds have a number of properties in common with TPA and TCDD. As shown in Table 1, each of these compounds is a potent skin irritant and rapidly induces alterations in patterns of epidermal cell growth and keratinization, and induces ornithine decarboxylase.[27,53,54] TPA and TCDD are tumor promoters in mouse skin, while the psoralens are suspected of being human skin tumor promoters as well as carcinogens.[55,56] In addition, the psoralens, like TPA and TCDD, also inhibit binding of EGF to responsive cell types (see below). Unlike TPA and TCDD, however, psoralens must be activated by UV light to exert biological effects. As indicated above, TPA and TCDD are thought to mediate their biological actions by binding to specific high-affinity receptors. It is our hypothesis that photoactivated psoralens also mediate their biological activity by binding to specific receptor sites. In support of this hypothesis, we have discovered specific, saturable, high-affinity binding sites for the psoralens that interact with the EGF receptor.[57,58] Indeed, the interaction of the psoralen receptors with the EGF receptor or other growth factor receptors may provide important clues as to the mechanism by which this class of chemicals initiates its biological activity in the skin. A more detailed description of our model on the mechanism of action of the psoralens is described at the end of this chapter.

It is important to note that not all of the activities of the psoralens, TCDD, and TPA are identical, and that the properties and functions of the receptors and target cells as well as the compounds themselves undoubtedly underlie differences. However, some of the bio-

Table 1

COMPARISON OF THE BIOLOGICAL PROPERTIES OF 8-MOP,
TCDD, AND TPA

Biological effect	8-MOP + UVA light	TCDD	TPA	Ref.
Skin toxicity	+	+	+	25,27,37,59
Cellular receptors	+	+	+	28,42,57
Active in low concentrations	+	+	+	28,42,57
Alterations in epidermal cell growth	+	+	+	12,32,33,39,60
Induction of ornithine decarboxylase	+	+	+	27,53,54
Alterations in keratinization	+	+	+	26,33,39,60
Inhibition of EGF binding	+	+	+	44—49,50—52,58
Activity at cell membrane	+	+	+	42,50—52,58,80
Tumor-promoting activity	?	+	+	37,41,55,56

Table 2

EXAMPLES OF IN VITRO CELLULAR MODELS USED
TO INVESTIGATE THE MECHANISM OF ACTION OF
THE PSORALENS AND UVA LIGHT

Cell type used	Biological effect	Ref.
Hamster ovary	Sister chromatid exchange	69,70
Human lymphocytes	Chromosome breakage	71
Human fibroblasts	Psoralen-DNA binding and cross-linking	75
Hamster fibroblasts	Mutagenicity	73,74
Human fibroblasts	DNA repair	67,68
Hamster fibroblasts	Cytotoxicity	72
Hamster fibroblasts	DNA synthesis	66

logical activities of these compounds are similar enough to suggest that, although the receptors for each compound are distinct, there must be overlap in the biochemical pathways activated following receptor binding. Thus, the principles used to explain the mechanism of action of TPA and TCDD are of value in understanding the biological responses observed with the psoralens.

Because of their biological activity as photosensitizing agents, the psoralens have been used as therapeutic agents in the skin. In humans, psoralen, when activated by ultraviolet (UVA, 320 to 400 nm) light (PUVA therapy), results in an inhibition of growth and maturation of epidermal keratinocytes and is used extensively in the photochemical treatment of epidermal proliferative disorders such as psoriasis.[59,60] PUVA also stimulates growth of melanocytes, as well as melanogenesis.[61] This process results in skin tanning and thus is effective in the treatment of vitiligo.[62] From these observations, it is apparent that photo-chemotherapy with the psoralens is associated with two clinically important, yet opposing effects on different epidermal cell populations. In order to understand the mechanism of action of the psoralens, these two opposing effects must be taken into account. A variety of in vitro mammalian cell models have been used to examine the mechanisms underlying psoralen phototoxicity (Table 2). Psoralens are known to bind covalently to DNA, and parameters such as the extent of psoralen-DNA binding and mono- and bifunctional adduct formation, as well as the effects of photoactivated psoralens on DNA synthesis and repair, sister chromatid exchange, chromosome breakage, and mutagenicity have been characterized.[63-75] Cell lines from many different sources have been used. Although inter-action with the DNA by photoactivated psoralens may be involved in some of the therapeutic responses to the drug, it is likely that other mechanisms also contribute to its actions. In

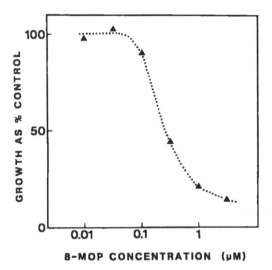

FIGURE 1.   Effects of 8-MOP and UVA light on the growth of HeLa cells. HeLa cells (5×10^4) in 3.5-cm culture dishes were incubated with increasing concentrations of 8-MOP for 30 min and then exposed to 0.69 J/cm^2 of UVA light as previously described.[57] Immediately following light treatment, the cells were refed with drug-free culture medium and grown in a humidified carbon dioxide incubator. After 4 days, the number of cells on the culture dishes was enumerated using a Coulter® counter. The relative number of cells on the plates at each drug concentration was compared to untreated controls. Neither 8-MOP nor UVA light alone had any effect on the growth of the cells. Each point represents the average of duplicate samples.

the sections below, we describe studies from our laboratory which demonstrate that, in mammalian cells, a site other than the DNA is a major target for the psoralens. This site is the psoralen receptor, and it has been localized in nonnuclear compartments of the cell. We have found that psoralen-receptor binding in cells in culture can be correlated with specific biological effects on the cell surface membrane, suggesting a role for this process in cell growth regulation.

II. INHIBITION OF CELL GROWTH IN VITRO BY PSORALENS AND UV LIGHT

In psoriasis, PUVA therapy leads to an inhibition of abnormally high rates of epidermal cell growth associated with the disease. We have determined that photoactivated psoralens are also potent inhibitors of epithelial cell growth in vitro. Using subconfluent cultures of the human carcinoma cell line HeLa, we found that treatment with the psoralen, 8-methoxypsoralen (8-MOP), followed by UV light exposure, produced a dose-dependent inhibition of cell proliferation (Figure 1). Growth inhibition occurred in the nanomolar psoralen concentration range over a period of 4 days. Neither UV light nor 8-MOP alone modulated growth of the cells. Since psoralen and UV light treatment in vitro appeared to mimic the biological effects of PUVA therapy in vivo, this cell culture model was used to further investigate the mechanism of action of the psoralens.

III. IDENTIFICATION OF PSORALEN RECEPTORS IN HeLa CELLS

In our initial studies, we used fluorescence microscopy to localize 8-MOP, an intrinsically fluorescent compound, in HeLa cells. We found that this psoralen was rapidly taken up by the cells. Interestingly, despite the fact that psoralens are known to intercalate into DNA,[65] we found that almost all of the 8-MOP associated with the cells was in the cytoplasm and cell surface membrane. Similar results have been reported previously by Moreno et al.[76] in cultured cells using microfluorimetry. Although both of these techniques underestimate DNA-associated psoralen due to fluorescence quenching, the data provided preliminary evidence that significant amounts of the psoralens were sequestered in sites other than the nucleus. Several other laboratories have also implicated targets outside of the nucleus for the actions of the psoralens, and these data are consistent with our observations.[77-81]

To characterize nonnuclear psoralen binding sites, we used a high specific activity radiolabeled [³H]-8-MOP.[57] In our initial studies, we monitored the kinetics of uptake of this psoralen into HeLa cells. When confluent cultures of cells were incubated with [³H]-8-MOP, we found that the radiolabeled compound was rapidly incorporated into the cells over time, reaching equilibrium within 30 min. The half-time of association of [³H]-8-MOP with the cells was less than 5 min at 4°C. A striking finding from these studies was that excess unlabeled 8-MOP significantly reduced binding of [³H]-8-MOP to the cells. These data suggested that the association of radiolabeled 8-MOP with HeLa cells could be saturated. Based on this observation, and on the rapid association of labeled 8-MOP with the cells, we proposed that psoralens bind to HeLa cells via specific saturable receptors and that these receptor sites are important in the biological activity of these compounds.[57] Several additional lines of evidence support this model. Receptor sites for 8-MOP in HeLa cells were found to be saturated with low concentrations of unlabeled psoralens. Furthermore, nonspecific binding of psoralens to the cells, that is, labeled 8-MOP binding that was not inhibited by excess unlabeled 8-MOP, was not saturable. In addition, binding of the radiolabeled 8-MOP to its receptor in HeLa cells was readily reversible. We also found that biologically active psoralen analogs including 4,5′,8-trimethylpsoralen (TMP) as well as psoralen itself were effective inhibitors of binding of labeled 8-MOP to its receptor (Figure 2). TMP was the most effective inhibitor, followed by 8-MOP and psoralen. The angular furocoumarin, 5-methylangelicin (5-MA), which is biologically inactive as a skin photosensitizer, was a poor inhibitor of 8-MOP receptor binding. This suggests that the psoralen receptor can readily discriminate between structurally related analogs. Other agents that have been used in the treatment of psoriasis, including methotrexate, benzoyl peroxide, anthralin, dexamethasone, and retinoic acid, were also ineffective inhibitors of 8-MOP receptor binding. These data indicate that psoralen binding to its receptor is also highly specific.

To further characterize psoralen binding sites, we next quantified saturation of psoralen receptor binding using the method of Scatchard.[82] From this analysis, we obtained a curvilinear graph with upward concavity (Figure 3). We interpreted this plot in terms of two independent psoralen binding sites with different affinities for 8-MOP. These were a higher-affinity site with a Kd of 19 nM and 1.8 × 10^5 receptor sites per cell and a lower-affinity site with a Kd of 4 μM and 7.1 × 10^6 receptor sites per cell. Thus, in addition to specificity, the psoralen receptor displays high- and low-affinity binding. These results are important since, as described above, many other growth factors, drugs, and hormones that modulate cellular proliferation are known to bind to specific high-affinity receptor sites in target cells.

HeLa is an epithelial cell line derived from a human cervical carcinoma. We also found that a variety of other epithelial-derived cells possess receptors for the psoralens, including the KB oral carcinoma, the B16 and G-361 melanomas, and the PAM 212 keratinocytes.[57] The fact that melanocyte- and keratinocyte-derived cells possess psoralen receptors is consistent with the known biological activity of the psoralens in the skin. In other studies, we

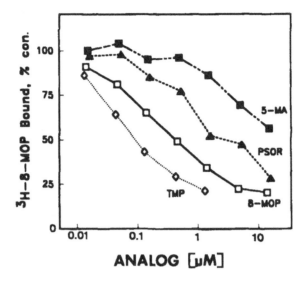

FIGURE 2. Ability of various psoralens to compete with labeled 8-MOP for receptor binding. Psoralen and several analogues, including 4,5′,8-trimethylpsoralen (TMP), 8-methoxypsoralen (8-MOP), and 5-methylangelicin (5-MA) were tested as inhibitors of [³H]-8-MOP receptor binding as previously described.[57] Inhibition of binding is presented as percentage of control binding in the absence of inhibitors. Each point represents the average of duplicate samples.

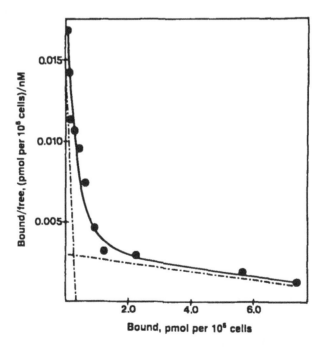

FIGURE 3. Scatchard plot of 8-MOP binding to HeLa cells. The plot was resolved into two linear components indicated by the dashed lines. (Reprinted from Laskin, J. D., Lee, E., Laskin, D., and Gallo, M. A., *Proc. Natl. Acad. Sci. U.S.A.*, 82, 6158, 1985. With permission.)

also noted that there were five- to tenfold fewer psoralen binding sites in fibroblast cell lines when compared to epithelial-derived cells.[57] This may explain the selective effects of PUVA treatment on the epithelial cells of the skin. It is also possible that the presence of high levels of psoralen receptors on cells represents a marker for epithelial cell development.

IV. UV LIGHT EXCITATION OF PSORALEN BOUND TO ITS RECEPTOR

The psoralens absorb UV light, and it is this property that confers biologic activity on these compounds. The biological action spectrum for 8-MOP lies in the range of 320 to 400 nm, with the greatest skin photosensitizing activity falling between 360 and 365 nm. Excitation of psoralen by 365-nm UV light causes the molecule to enter into photoreactive electronic excited states.[65] Once inside the cells, reactive psoralen intermediates formed during this process may combine with molecular oxygen to form potentially toxic reactive oxygen species, including superoxide anion. Photoactivated psoralens may also react with specific cellular components, forming adducts with DNA, RNA, protein, and phospholipids.[65,81,83]

One of the most well-studied targets of psoralen is the DNA. Psoralens intercalate into DNA in a dark reaction. Following UV light activation, they form mono- and bifunctional adducts with the pyrimidine bases in DNA.[65,81] Since we found that 8-MOP also binds to specific high-affinity receptor sites, it was of interest to determine if these sites were also covalently modified by the psoralen following treatment with UV light. Equilibrium binding of [3H]-8-MOP to its receptor is a kinetic phenomenon and does not require light treatment. However, once bound to its receptor in cells and activated by UV light, we found that [3H]-8-MOP could form covalent modifications. Covalent modification of the receptors was inhibited by excess unlabeled 8-MOP, indicating that this type of receptor binding was also saturable. Interestingly, in HeLa cells treated with [3H]-8-MOP and UV light, only a small proportion (<3%) of the receptors binding the drug actually became covalently modified. It is possible that this represents only the population of higher-affinity psoralen receptors. The role of covalent modification of the receptor in biological activity remains to be determined. If the process is critical, then only a very small number of activated receptors appear to be required for activity of the psoralens. Alternatively, one could speculate that it is the photoreactive electronic excited state of the molecule, and not covalent modifications, that actually activates the psoralen receptor and initiates biological activity. Further studies are necessary to explore this possibility.

V. CHARACTERIZATION OF THE PSORALEN RECEPTOR

There are several lines of evidence to indicate that the psoralen receptor is not associated with the DNA. As described above, studies on the distribution of the compounds in cells based on their fluorescence properties demonstrated that psoralen was present in the cytoplasm and cell surface membranes. In addition, when cell membrane fractions of HeLa cells were prepared, psoralen receptor binding could be detected. Taken together, these data indicate that at least some of the psoralen receptor sites are located in cell membranes. Using [3H]-8-MOP, we also examined the subcellular distribution of covalently bound psoralen in HeLa cells. In these experiments, cells were treated with [3H]-8-MOP and then pulsed with UV light. We found that the labeled 8-MOP was distributed in nuclear as well as cytoplasmic and membrane fractions of the cells.[84] When the HeLa cells were treated with [3H]-8-MOP in the presence of an excess of unlabeled 8-MOP and then exposed to UV light, there was dramatically less covalent binding of the psoralen analog to the cytoplasmic and membrane fractions of the cells, that is, psoralen photobinding was saturated. To date, we have been unable to demonstrate a similar saturation of covalent binding of psoralen to DNA.[57] In further studies, we found that the psoralen-receptor complex was resistant to DNAse and RNAse treatment, but sensitive to proteases.[84] These data provide support for our model that psoralen receptor binding is not associated with the DNA and indicate that, in fact, the psoralen receptor may be a protein.

By means of sodium dodecyl sulfate polyacrylamide gel electrophoresis and fluorography, the psoralen receptor was further characterized in HeLa cell extracts containing radiolabeled 8-MOP covalently bound to its receptor. The photoalkylated psoralen receptor was found to migrate in the gels as a protein with a molecular mass of approximately 22,000 daltons.[83] Unlabeled 8-MOP completely prevented binding of the radiolabel to this receptor, providing further evidence that psoralen receptor binding is saturable. In these studies, the receptor was localized exclusively in cytoplasmic and membrane fractions of HeLa cells. These findings suggest that receptor-mediated biological activity of the psoralens does not involve direct interaction with DNA.

VI. INHIBITION OF EPIDERMAL GROWTH FACTOR BINDING BY PHOTOACTIVATED PSORALENS

The fact that the photoactivated psoralens are potent modulators of epidermal cell growth prompted us to investigate whether these compounds interact with growth factors known to regulate epidermal cell proliferation. We chose to examine EGF, since its effects on mammalian cell growth are well characterized.[1-4] EGF is a low-molecular-weight polypeptide that binds to cell surface receptor sites on responsive cell types. The EGF receptor is a tyrosine-specific protein kinase whose activity is augmented following EGF binding.[4] Once EGF binds to and activates its receptor, the EGF-receptor complex is internalized by the cell in a temperature-dependent process via the clathrin-coated pit pathway.[2,4] When confluent cultures of HeLa cells were treated with 8-MOP and UV light, we observed a decrease in binding of radiolabeled EGF to its receptor (Figure 4).[58] The effects of PUVA on EGF binding were in general comparable to those observed following treatment of cells with the phorbol ester TPA, a well-characterized inhibitor of EGF binding. The 8-MOP-induced inhibition of EGF binding to HeLa cells was very rapid, occurring immediately following PUVA treatment, and required UV light. Several biologically active psoralen analogs, including 5-methoxypsoralen (5-MOP), TMP, and psoralen, when activated by UV light were also effective inhibitors of EGF binding.[58] In addition, PUVA treatment was found to inhibit EGF binding to a variety different mouse and human cells lines known to possess psoralen receptors, including those derived from the epidermis. These data indicate that the ability of photoactivated psoralens to interact with the EGF receptor and modulate EGF binding may be a characteristic property of these compounds. The fact that the psoralens are effective in epidermal-derived cells is important, since this cell type is the major target for these drugs.

In further studies we found that inhibition of EGF binding by psoralens was not due to direct competition with EGF for its cell surface receptor.[58] In addition, while PUVA inhibited EGF binding at 37°C, it had no effect on binding at 4°C[85] or on EGF binding in broken cell preparations. Taken together, these data suggest that cellular metabolic activity is required for the biological activity of PUVA and that inhibition of EGF binding by this treatment probably occurs by an indirect mechanism. The tumor promoter TPA is also known to inhibit EGF binding through an indirect mechanism.[44,47]

VII. A MODEL FOR THE MECHANISM OF ACTION OF THE PSORALENS

Our findings that mammalian cells possess specific receptors for the psoralens and that different psoralen analogs bind to these receptors in a manner that parallels their biological

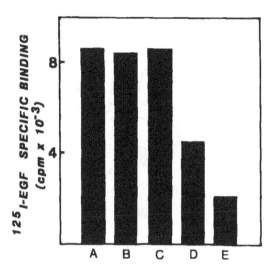

FIGURE 4. Inhibition of ^{125}I-EGF specific binding in HeLa cells by 8-MOP and UVA light. The specific binding of ^{125}I-EGF to HeLa cells was determined as previously described.[58] The data, presented as the amount of ^{125}I-EGF receptor binding in 3×10^6 cells, is the average of duplicate samples. (A) Untreated control cells; (B) cells treated with UVA light alone (3.7 J/cm²); (C) cells treated with 8-MOP (4.4 μM); (D) cells treated with 8-MOP (4.4 μM) and UVA light (3.7 J/cm²); (E) cells treated with TPA (162 nM). Cells were pretreated with 8-MOP and TPA for 30 min and then assayed for EGF binding at 37°C. 8-MOP-treated cells were exposed to UVA light immediately following the pretreatment period.

activity in skin photosensitization assays suggest that these receptors play a role in cell growth regulation. Based on these observations, we have developed a model for psoralen action (Figure 5). According to our model, psoralens bind to specific receptors on responsive cells. Exposure to UV light induces psoralen receptor activation and initiates intracellular signals that lead to biological responses. It is our hypothesis that these signals involve interaction with normal growth factor receptors such as EGF. At this time we cannot rule out the possibility that psoralen binding to other cellular components also contributes to the biological effects of PUVA.

EGF modulates a variety of cellular activities, including nutrient transport; phospholipid turnover; protein phosphorylation; endocytosis; glycolysis; ornithine decarboxylase; and DNA, RNA, and protein synthesis.[1-3] The effects of EGF on these processes are initiated by regulatory signals generated from the activation of the EGF receptors and its associated tyrosine kinase activity. The cellular response to these regulatory signals is dependent on the origin of the cell. For example, EGF inhibits growth of A431 epidermoid cells[86] and stimulates the growth of human fibroblasts.[32] The photoactivated psoralens, by inhibiting the binding of growth factors, may in turn also exert growth promoting as well as growth inhibitory activity. Our model would thus explain the opposing effects of PUVA on the growth of different epidermal cell populations.

As outlined in the introduction, both TPA and TCDD modulate cell growth. In fact, like the psoralens, these compounds also inhibit, as well as promote, the growth of different cell types. Inhibition of EGF binding in cells by TPA and psoralens occurs in a nearly identical manner, suggesting that their mechanism of action might be similar.[48,49,58] However, these

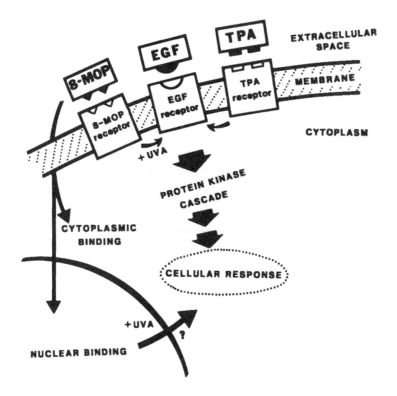

FIGURE 5. Model for the mechanism of action of the psoralens. The psoralens may interact with their receptors as well as other components in the cells, including the DNA. In this model, both the 8-MOP and TPA receptors may be localized in the cell cytoplasm. UVA light activates the 8-MOP receptor and possibly other intracellular binding sites, initiating biological responses. Activation of the TPA receptor does not require UVA light. The activated TPA and 8-MOP receptors interact with the EGF receptor to modulate EGF binding and its protein kinase activity, thus altering the cell's response to a growth factor stimulus.

compounds appear to act by binding to and activating distinct receptors. It has been shown that TPA binds to the calcium- and phospholipid-dependent protein kinase, protein kinase C.[42] Once activated, protein kinase C translocates from the cytoplasm to the cell surface membrane, where it presumably phosphorylates critical proteins involved in cell growth regulation. The EGF receptor has been shown to be a substrate for protein kinase C. Phosphorylation of the EGF receptor by protein kinase C is associated with decreased EGF receptor tyrosine kinase activity as well as EGF binding.[87,88]

Based on our model, we propose that the photoactivated psoralen receptor, like the TPA receptor, modulates the EGF receptor by inducing its phosphorylation. One could speculate that this may occur indirectly, through the production of diacylglycerol, the endogenous activator of protein kinase C. Further studies are necessary to explore the potential role of protein kinase C in the action of the psoralens and the role of the psoralen receptor in cell growth regulation.

VIII. SUMMARY AND CONCLUSIONS

The psoralens have been gaining widespread acceptance as therapeutic agents for a number of diverse skin-related diseases. Based on the many biological responses observed with these compounds, it appears that their interaction with DNA alone is not sufficient to explain their mechanism of action. Our laboratory has characterized specific high-affinity receptor sites for the psoralens in responsive cell types, and we have hypothesized that these receptors mediate, at least in part, the biological actions of these compounds. One mechanism by which this may occur is through an interaction with a normal growth factor receptor in the skin. We have demonstrated that one potential receptor is EGF. Photoactivated psoralens inhibit binding of EGF. These data demonstrate that the psoralens can have biological effects on the cell surface membrane. Characterization of the psoralen receptor in different skin cells and the events that occur following receptor activation are necessary to understand how PUVA treatment affects the EGF receptor and subsequently regulates cell growth in diseases such as psoriasis.

ACKNOWLEDGMENTS

This work was supported by grant ES-03647 from the National Institute of Environmental Health Sciences. The authors are indebted to Dr. Fredika M. Robertson for critically reading this manuscript.

REFERENCES

1. **Carpenter, G. and Cohen, S.,** Epidermal growth factor, *Annu. Rev. Biochem.,* 48, 193, 1979.
2. **Das, M.,** Epidermal growth factor: mechanisms of action, *Int. Rev. Cytol.,* 78, 233, 1982.
3. **Carpenter, G. and Cohen, S.,** Human epidermal growth factor and the proliferation of human fibroblasts, *J. Cell. Physiol.,* 71, 227, 1976.
4. **Carpenter, G.,** Properties of the receptor for epidermal growth factor, *Cell,* 37, 357, 1984.
5. **Ross, R., Raines, E. W., and D. F. Bowen-Pope,** The biology of platelet-derived growth factor, *Cell,* 46, 155, 1986.
6. **Baird, A., Frederick, E., Mormede, P., Ueno, N., Bohlen, P., Ying, S.-Y., Wehrenberg, W. B., and Guillemin, R.,** Molecular characterization of fibroblast growth factor: distribution and biological activities in various tissues, *Recent Prog. Horm. Res.,* 42, 143, 1986.
7. **Huang, J. S., Huang, S. S., and Kuo, M.-D.,** Bovine brain-derived growth factor: purification and characterization of its interaction with responsive cells, *J. Biol. Chem.,* 261, 11, 600, 1986.
8. **Gospodarowicz, D., Greenburg, G., Bialecki, H., and Zetter, B. R.,** Factors involved in the modulation of cell proliferation *in vivo* and *in vitro:* the role of fibroblasts and epidermal growth factor in the proliferative response of mammalian cells, *In Vitro,* 14, 85, 1978.
9. **Roberts, A. B., Anzano, M. A., Wakefield, L. M., Roche, N. S., Stern, D. F., and Sporn, M. B.,** Type beta transforming growth factor: a bifunctional regulator of cellular growth, *Proc. Natl. Acad. Sci. U.S.A.,* 82, 119, 1985.
10. **Tucker, R. F., Branum, E. L., Shipley, G. D., Ryan, R. J., and Moses, H. L.,** Specific binding to cultured cells of ^{125}I-labeled type beta transforming growth factor from human platelets, *Proc. Natl. Acad. Sci. U.S.A.,* 81, 6757, 1984.
11. **Frolik, C. A., Wakefield, L. M., Smith, D. M., and Sporn, M. B.,** Characterization of a membrane receptor for transforming growth factor in normal rat kidney fibroblasts, *J. Biol. Chem.,* 259, 10995, 1984.
12. **Rice, R. H. and Cline, P. R.,** Opposing effects of 2,3,7,8-tetrachlorodibenzo-*p*-dioxin and hydrocortisone on growth and differentiation of cultured malignant keratinocytes, *Carcinogenesis,* 5, 367, 1984.
13. **Kalimi, M., Beato, M., and Feigelson, P.,** Interaction of glucocorticoids with rat liver nuclei, I. Role of cytosol proteins, *Biochemistry,* 12, 3365, 1973.
14. **Baxter, J. D. and Tomkins, G. M.,** Specific cytoplasmic glucocorticoid receptors in hepatoma tissue culture cells, *Proc. Natl. Acad. Sci. U.S.A.,* 68, 932, 1971.
15. **Beato, M. and Feigelson, P.,** Glucocorticoid-binding proteins of rat liver cytosol. I. Separation and identification of the binding properties, *J. Biol. Chem.,* 247, 7890, 1972.
16. **Fisher, L. B. and Maibach, H. I.,** The effect of corticosteroids on human epidermal mitotic activity, *Arch. Dermatol.,* 103, 39, 1971.
17. **Levine, L.,** Arachidonic acid transformation and tumor production, *Adv. Cancer Res.,* 35, 49, 1981.
18. **Honn, K. V. and Meyer, J.,** Thromboxanes and prostacyclin: positive and negative modulators of tumor growth, *Biochem. Biophys. Res. Commun.,* 102, 1122, 1981.

19. **Honn, K. V., Bockman, R. S., and Marnett, L. J.,** Prostaglandins and cancer: a review of tumor initiation through tumor metastasis, *Prostaglandins*, 21, 833, 1982.

20. **Ford-Hutchinson, A. W. and Chan, C. C.,** Pharmacological actions of leukotrienes in the skin, *Br. J. Dermatol.*, 113, 95, 1985.

21. **Kragballe, K. and Vorhees, J. J.,** Leukotrienes are potent stimulators of DNA synthesis of human keratinocyte cultures, *J. Invest. Dermatol.*, 82, 398, 1984.

22. **Jetten, A. M.,** Actions of retinoids and phorbol esters on growth and binding of epidermal growth factor, *Ann. N.Y. Acad. Sci.*, 359, 200, 1981.

23. **Jetten, A. M. and Goldfarb, R. H.,** Action of epidermal growth factor and retinoids on anchorage-dependent and -independent growth of non-transformed rat kidney cells, *Cancer Res.*, 43, 2094, 1983.

24. **Roberts, A. and Sporn, M. B.,** Cellular biology and biochemistry of the retinoids, in *The Retinoids*, Vol. 2, Sporn, M. B., Roberts, A. B., and Goodman, D. S., Eds., Academic Press, Orlando, Fla., 1984, 209.

25. **Kimbrough, R. D.,** The toxicity of polychlorinated polycyclic compounds and related chemicals, *Crit. Rev. Toxicol.*, 2, 445, 1974.

26. **Knutson, J. and Poland, A.,** Keratinization of mouse teratoma cell line XB produced by 2,3,7,8-tetrachlorodibenzo-*p*-dioxin: an in vitro model of toxicity, *Cell*, 22, 27, 1980.

27. **Knutson, J. C. and Poland, A.,** Response of murine epidermis to 2,3,7,8-tetrachlorodibenzo-*p*-dioxin: interaction of the Ah and hr loci, *Cell*, 30, 225, 1982.

28. **Poland, A. and Knutson, J. C.,** 2,3,7,8-Tetrachlorodibenzo-*p*-dioxin and related aromatic hydrocarbons: examination of the mechanism of toxicity, *Annu. Rev. Pharmacol. Toxicol.*, 22, 517, 1982.

29. **Greenlee, W. F., Osborne, R., Hudson, L. G., and Toscano, W. A.,** Studies on the mechanisms of toxicity of TCDD to human epidermis, in *Banbury Report 18, Biological Mechanisms of Dioxin Action*, Poland, A. and Kimbrough, R. D., Eds., Cold Spring Harbor Laboratory, New York, 1984, 365.

30. **Rice, R. H. and P. R. Cline,** Response of malignant epidermal keratinocytes to 2,3,7,8-TCDD, in *Banbury Report 18, Biological Mechanisms of Dioxin Action*, Poland, A. and Kimbrough, R. D., Eds., Cold Spring Harbor Laboratory, New York, 1984, 373.

31. **Poland, A.,** Reflections on the mechanism of action of halogenated aromatic hydrocarbons, in *Banbury Report 18, Biological Mechanisms of Dioxin Action*, Poland, A. and Kimbrough, R. D., Eds., Cold Spring Harbor Laboratory, New York, 1984, 109.

32. **Milstone, L. and LaVigne, J.,** 2,3,7,8-Tetrachlorodibenzo-*p*-dioxin induces hyperplasia in confluent cultures of human keratinocytes, *J. Invest. Dermatol.*, 82, 532, 1984.

33. **Molloy, C. J. and Laskin, J. D.,** Alterations in the expression of specific epidermal keratin markers in the hairless mouse by the topical application of the tumor promoters 2,3,7,8-tetrachlorodibenzo-*p*-dioxin (TCDD) and the phorbol ester 12-*O*-tetradecanoylphorbol-13-acetate (TPA), *Carcinogenesis*, 8, 1193, 1987.

34. **Crow, K. D.,** Chloracne (halogen acne), in *Dermatology*, 2nd ed., Marzulli, F. N. and Maibach, H. T., Eds., Hemisphere, New York, 1983, 461.

35. **Puhvel, S. M., Sakamoto, M., and Reisner, R. M.,** Hairless mice as models for chloracne: a study of cutaneous changes induced by topical application of established chloracnegens, *Toxicol. Appl. Pharmacol.*, 64, 492, 1982.

36. **Boutwell, R. K.,** The function and mechanism of promoters of carcinogenesis, *Crit. Rev. Toxicol.*, 2, 419, 1974.

37. **Berenblum, I.,** Sequential aspects of chemical carcinogenesis: skin, in *Cancer: A Comprehensive Treatise*, Vol. 1, Becker, F. F., Ed., Plenum Press, New York, 1975, 324.

38. **Raick, A. N.,** Ultrastructural, histological and biochemical alteration produced by 12-*O*-tetradecanoylphorbol-13-acetate on mouse epidermis and their relevance to skin tumor promotion, *Cancer Res.*, 33, 269, 1973.

39. **Molloy, C. J. and Laskin, J. D.,** Specific alterations in keratin biosynthesis in mouse epidermis in vivo and in explant culture following a single exposure to the tumor promoter 12-*O*-tetradecanoylphorbol-13-acetate, *Cancer Res.*, 47, 4674, 1987.

40. **Laskin, J. D., Mufson, R. A., Piccinini, L. Engelhardt, D. L., and Weinstein, I. B.,** Effects of the tumor promoter 12-*O*-tetradecanoylphorbol-13-acetate on newly synthesized proteins in mouse epidermis, *Cell*, 25, 441, 1981.

41. **Poland, A., Palen, D., and Glover, E.,** Tumor promotion by TCDD in skin of HRS/J hairless mice, *Nature (London)*, 300, 271, 1982.

42. **Nishizuka, Y.,** The role of protein kinase C in cell surface signal transduction and tumor promotion, *Nature (London)*, 308, 693, 1984.

43. **Cuthill, S., Poellinger, L., and Gustafsson, J. A.,** The receptor for 2,3,7,8-tetrachlorodibenzo-*p*-dioxin in the mouse hepatoma cell line Hepa 1c1c7: a comparison with the glucocorticoid receptor and the mouse and rat hepatic dioxin receptors, *J. Biol. Chem.*, 262, 3477, 1987.

44. **Lee, L.-S. and Weinstein, I. B.,** Tumor-promoting phorbol esters inhibit binding of epidermal growth factor to cellular receptors, *Science*, 202, 313, 1978.

45. **Lee, L.-S. and Weinstein, I. B.,** Mechanism of tumor promoter inhibition of cellular binding of epidermal growth factor, *Proc. Natl. Acad. Sci. U.S.A.,* 76, 5168, 1979.
46. **Shoyab, M., De Larco, J. E., and Todaro, G. J.,** Biologically active phorbol esters specifically alter the affinity of epidermal growth factor membrane receptor, *Nature (London),* 279, 387, 1979.
47. **Brown, K. D., Dicker, P., and Rozengurt, E.,** Inhibition of epidermal growth factor binding to surface receptors by tumor promoters, *Biochem. Biophys. Res. Commun.,* 86, 1037, 1979.
48. **Magun, B. E., Matrisian, L. M., and Bowden, G. T.,** Epidermal growth factor: ability of tumor promoter to alter its degradation, receptor affinity and receptor number, *J. Biol. Chem.,* 255, 6373, 1980.
49. **Lockyer, J. M., Bowden, G. T., Matrisian, L. M., and Magun, B. E.,** Tumor promoter-induced inhibition of epidermal growth factor binding in cultured mouse primary epidermal cells, *Cancer Res.,* 41, 2308, 1981.
50. **Madhukar, B. V., Brewster, D. W., and Matsumura, F.,** Effect of *in vivo*-administered 2,3,7,8-tetrachlorodibenzo-*p*-dioxin on receptor binding of epidermal growth factor in the hepatic plasma membranes of rat, guinea pig, mouse and hamster, *Proc. Natl. Acad. Sci. U.S.A.,* 81, 7407, 1984.
51. **Matsumura, F., Madhukar, B. V., Bombick, D. W., and Brewster, D. W,** Toxicological significance of pleiotropic changes of plasma membrane functions particularly that of EGF receptor caused by 2,3,7,8-TCDD, in *Banbury Report 18, Biological Mechanisms of Dioxin Action,* Poland, A. and Kimbrough, R. D., Eds., Cold Spring Harbor Laboratory, New York, 1984, 267.
52. **Hudson, L. G., Toscano, W. A., and Greenlee, W. F.,** Regulation of epidermal growth factor binding in a human keratinocyte cell line by 2,3,7,8-tetrachlorodibenzo-*p*-dioxin, *Toxicol. Appl. Pharmacol.,* 77, 251, 1985.
53. **Lowe, N. J., Connor, M. J., Cheong, E. S., Akopsantz, and Breeding, J. H.** Psoralen and ultraviolet light. A. Effects on epidermal ornithine decarboxylase induction and DNA synthesis in the hairless mouse, *Natl. Cancer Inst. Monogr.,* 66, 73, 1984.
54. **O'Brien, T. G.,** The induction of ornithine decarboxylase as an early, possibly obligatory event in mouse skin carcinogenesis, *Cancer Res.,* 36, 2644, 1976.
55. **Stern, R. S., Thibodeau, L. A., Kleinerman, R. A., Parrish, J. A., Fitzpatrick, T. B., and 22 participating investigators,** Risk of cutaneous carcinoma in patients treated with oral methoxsalen photochemotherapy for psorisis, *N. Engl. J. Med.,* 300, 809, 1979.
56. **Stern, R. S., Laird, N., Melski, J., Parrish, J. A., Fitzpatrick, T. B., and Bleich, H. L.,** Cutaneous squamous-cell carcinoma in patients treated with PUVA, *N. Engl. J. Med.,* 310, 1156, 1984.
57. **Laskin, J. D., Lee, E., Yurkow, E. J., Laskin, D. L., and Gallo., M. A.,** A possible mechanism of psoralen phototoxicity not involving direct interaction with DNA, *Proc. Natl. Acad. Sci. U.S.A.,* 82, 6158, 1985.
58. **Laskin, J. D., Lee, E., Laskin, D., and Gallo, M. A.,** Psoralens potentiate ultraviolet-induced inhibition of epidermal growth factor binding, *Proc. Natl. Acad. Sci. U.S.A.,* 83, 8211, 1986.
59. **Gange, R. W. and Parrish, J. A.,** Cutaneous toxicity due to psoralens, *Natl. Cancer Inst. Monogr.,* 66, 117, 1984.
60. **Fitzpatrick, T. B. and Pathak, M. A.,** Research and development of oral psoralen and longwave radiation photochemotherapy: 2000 B.C.—1982 A.D., *Natl. Cancer Inst. Monogr.,* 66, 3, 1984.
61. **Pathak, M. A., Kramer, D. M., and Fitzpatrick, T. B.,** Photobiology and photochemistry of furocoumarins (psoralens), in *Sunlight and Man: Normal and Abnormal Photobiological Responses,* Fitzpatrick, T. B., Pathak, M. A., Haber, L. C., Seiji, M., and Kukita, A., Eds., University of Tokyo Press, Tokyo, 1974, 335.
62. **Nordlund, J. and Lerner, A. B.,** Vitiligo: it is important, *Arch. Dermatol.,* 118, 5, 1982.
63. **Grosswiener, L. I.,** Mechanism of photosensitization by furocoumarins, *Natl. Cancer Inst. Monogr.,* 66, 47, 1984.
64. **Pathak, M. A.,** Mechanisms of psoralen photosensitization reactions, *Natl. Cancer Inst. Monogr.,* 66, 41, 1984.
65. **Song, P. S. and Tapley, K. J.,** Photochemistry and photobiology of psoralens, *Photochem. Photobiol.,* 29, 1177, 1979.
66. **Trosko, J. E. and Isoun, M.,** Photosensitizing effect of trisoralen on DNA synthesis in human cells grown in vitro, *Int. J. Radiat. Biol.,* 19, 87, 1971.
67. **Gruenert, D. C. and Cleaver, J. E.,** Repair of psoralen-induced cross-links and mono-adducts in normal and repair deficient human fibroblasts, *Cancer Res.,* 45, 5399, 1985.
68. **Zolan, M. E., Smith, C. A., and Hanawalt, P. C.,** Repair of furocoumarin adducts in mammalian cells, *Natl. Cancer Inst. Monogr.,* 66, 137, 1984.
69. **Cassel, D. M. and Latt, S. A.,** Relationship between DNA adduct formation and sister chromatid exchange induction by (^{3}H) 8-methoxypsoralen in Chinese hamster ovary cells, *Exp. Cell Res.,* 128, 15, 1980.
70. **Loveday, K. S. and Donahue, B. A.,** Induction of sister chromatid exchanges and gene mutations in Chinese hamster ovary cells by psoralens, *Natl. Cancer Inst. Monogr.,* 66, 149, 1984.

71. **Sasaki, M. S. and Tonomura, A.**, A high susceptibility of Fanconi's anemia to chromosome breakage by DNA cross-linking agents, *Cancer Res.*, 1829, 1973.

72. **Ben-Hur, E. and Elkind, M. M.**, Psoralen plus near ultraviolet light inactivation of cultured Chinese hamster cells and its relation to DNA cross-links, *Mutat. Res.*, 18, 315, 1973.

73. **Averback, D., Papadopoulo, D., and Quinto, I.**, Mutagenic effects of psoralens in yeast and V79 Chinese hamster cells, *Natl. Cancer Inst. Monogr.*, 66, 127, 1984.

74. **Babudri, N., Pani, B., Venturini, S., Tamaro, M., Monti-Bragadin, C., and Bordin, F.**, Mutation induction and killing of V79 Chinese hamster cells by 8-methoxypsoralen plus near-ultraviolet light: relative effects of monoadducts and crosslinks, *Mutat. Res.*, 91, 391, 1981.

75. **Gruenert, D. C., Ashwood-Smith, M., Mitchell, R. H., and Cleaver, J. E.**, Induction of DNA-DNA cross-link formation in human cells by various psoralen derivatives, *Cancer Res.*, 45, 5394, 1985.

76. **Moreno, G., Salet, C., Kohen, C., and Kohen, E.**, Penetration and localization of furocoumarins in single living cells studied by microspectrofluorometry, *Biochem. Biophys. Acta*, 721, 109, 1982.

77. **Pathak, M. A. and Kramer, D. M.**, Photosensitization of skin in vivo by furocoumarins (psoralens), *Biochim. Biophys. Acta*, 195, 197, 1969.

78. **Frederichsen, S. and Hearst, J. E.**, Binding of 4'-amino-4,5',8-trimethylpsoralen to DNA, RNA and protein in HeLa cells and *Drosophila* cells, *Biochim. Biophys. Acta*, 563, 343, 1979.

79. **Bertaux, B., Dubertret, L., and Moreno, G.**, Autoradiographic localization of 8-methoxypsoralen in psoriasis skin *in vitro*, *Acta Derm. Venereol. (Stockholm)*, 61, 481, 1981.

80. **Toda, K., Danno, K., Tachibana, T., and Horio, T.**, Effect of 8-methoxypsoralen plus long wave ultraviolet (PUVA) radiation on mast cells. II. *In vitro* PUVA inhibits degranulation of rat peritoneal mast cells induced by compound 48/80, *J. Invest. Dermatol.*, 87, 113, 1986.

81. **Ben-Hur, E. and Song, P. S.**, The photochemistry and photobiology of furocoumarins, *Adv. Radiat. Biol.*, 11, 131, 1984.

82. **Scatchard, G.**, The attraction of proteins for small molecules and ions, *Ann. N.Y. Acad. Sci.*, 51, 660, 1949.

83. **Kitten, L., Midden, W. R., and Wang, S. Y.**, Photoreactions of lipids sensitized by furocoumarins, *Photochem. Photobiol.*, 37S, 16, 1983.

84. **Yurkow, E. J. and Laskin, J. D.**, Characterization of a photoalkylated psoralen receptor in HeLa cells, *J. Biol. Chem.*, 262, 8439, 1987.

85. **Laskin, J. D. and Lee, E.**, Effect of photoactivated psoralens on EGF binding in KB cells, submitted.

86. **Barnes, D.**, Epidermal growth factor inhibits growth of A431 human epidermoid carcinoma in serum-free culture medium, *J. Cell Biol.*, 93, 1, 1982.

87. **Cochet, C., Gill, G. N., Meisenhelder, J., Cooper, J. A., and Hunter, T. A.**, C-kinase phosphorylates the EGF receptor and reduces its epidermal growth factor-stimulated tyrosine protein kinase activity, *J. Biol. Chem.*, 259, 2533, 1984.

88. **Friedman, B., Frackelton, A. R., Ross, A. H., Conners, J. M., Fujiki, H., Sugimura, T., and Rosner, M. R.**, Tumor promoters block tyrosine specific phosphorylation of the epidermal growth factor receptor, *Proc. Natl. Acad. Sci. U.S.A.*, 81, 3034, 1984.

INDEX

Q